Understanding Systems

A Grand Challenge for 21st Century Engineering

Understanding Systems

A Grand Challenge for 21st Century Engineering

Jamshid Ghaboussi
Michael F Insana
University of Illinois at Urbana-Champaign, USA

World Scientific

NEW JERSEY · LONDON · SINGAPORE · BEIJING · SHANGHAI · HONG KONG · TAIPEI · CHENNAI · TOKYO

Published by

World Scientific Publishing Co. Pte. Ltd.

5 Toh Tuck Link, Singapore 596224

USA office: 27 Warren Street, Suite 401-402, Hackensack, NJ 07601

UK office: 57 Shelton Street, Covent Garden, London WC2H 9HE

British Library Cataloguing-in-Publication Data
A catalogue record for this book is available from the British Library.

UNDERSTANDING SYSTEMS
A Grand Challenge for 21st Century Engineering

ISBN 978-981-3225-94-7
ISBN 978-981-3225-95-4 (pbk)

Desk Editor: Christopher Teo

Typeset by Stallion Press
Email: enquiries@stallionpress.com

To our families

Preface

This text is aimed at improving reader intuition about the behavior of simple and *complex systems*. The target audience includes engineering instructors and students curious to explore the grand challenges on which careers are selected.

In general, systems are assemblies of intercommunicating components that we find throughout nature and society. They are of primary interest because of the significant influence they have on our daily life. Complexity is an important feature of some systems that is difficult to precisely define, but is often easy to recognize from characteristic properties such as *emergence* and sudden equilibrium-state *transitions*. From solid-state physics to insect colonies, large corporations and social networks each field of study recognizes complexity in their analysis precisely because they share common properties. Defining complex systems is a little like walking in the woods at night; properties are difficult to see if you look at each component directly, but are more readily sensed by our peripheral vision as large-scale properties emerge. How can something so important and fundamental in nature and society still be poorly understood?

Part of the answer lies in the difficulty of objectively seeing any environment in which one is totally immersed. Consider for example the round-versus-flat-Earth debate. Before the 3rd century BC, ancient Greek scholars knew the Earth was round by studying clues offered by the moon and eclipses. Yet debate raged for many centuries until Magellan circumnavigated the globe in the 16th century. The evidence became plain for all in the 1960s during the moon missions when those amazing images of our round blue globe first were published. Often new insights are found from changing our perspective on a problem, and that, in part, is what we hope to achieve here with regard to understanding systems.

Eliminating confusion about complexity also requires that we have available the analytical tools, shared experiences, and a descriptive language necessary to understand the many different abstract concepts associated with the nature of systems. For many in science and engineering, the tools of modeling and the language of mathematics hold the key to understanding any and all phenomena, even those as complex as the origins of life on Earth. Reductionist tools, which have served us well for hundreds of years when we hope to explain nature's mechanisms, yield few insights when studying the largest and most complex systems. Mathematics is a tool for understanding when using our mind's foveal vision. Nevertheless, the literature is full of attempts to apply mathematical modeling for the analysis of complex systems. We can find numerous texts describing properties of linear and nonlinear systems, and the roles of feedback, stability, chaos, control, robustness, or the tools provided by information, graph and bifurcation theories. Each offers credible insights into system *mechanisms* that have advanced our ability to predict aspects of system behavior, but often those views are simplified. The full nature of complex systems continues to escape our collective understanding, and, more importantly, our ability to predict their volatile behavior.

With this in mind, the authors have been meeting for at least an hour each week for nearly five years to explore alternative perspectives and approaches to systems analysis. We think we have a promising direction identified but, because it is still viewed through our peripheral vision, it has not yet risen to the level of a mathematical theory. (At least one of us doubts it ever will!) Consequently the story is difficult to tell. On the other hand, our story does borrow a generalized version of an important mathematical tool from engineering - eigenanalysis - to explore complex systems. Unfortunately this tool comes with baggage. Knowledgeable engineers and scientists say, "Wait a minute! Eigenanalysis only describes linear systems and all of the most interesting systems are nonlinear." True. Also, most advanced engineering students can compute eigenvalues when given a matrix but they have little or no intuition about what they mean physically and how essential they are to the nature of a system. We feel it is important to explore complex systems without mathematical analyses, at least initially, because math can push us into thinking traps that prevent new ideas from emerging. A technical but math-free description of systems is our goal in this book.

We set out to tell a very technical story using only shared concepts from everyday life and classical problems well known from a reader's schooling or reading in science and maybe a little engineering. With something as complicated as complex systems, intuition is a much more powerful tool for initial exploration than the analytical skills needed for advanced research. We consider the following chapters as an unusual type of engineering textbook. One intended to both challenge the assumptions of classically-trained engineers and introduce important new concepts to students just beginning to think about systems. We claim no confirmed breakthroughs in understanding specific complex system, but we hope the discussion provides interested readers with new intuition about all complex systems based on fundamental principles that shed light on their inner workings. These tools include an expanded description of eigenanalysis beyond the usual story from linear algebra, and suggestions for new modeling methods that can help practicing engineers explore the phenomena known as complexity.

Why would anyone write a book that poses questions without also offering verified solutions? The answer, of course, is to excite and prepare capable minds to engage the biggest questions of 21st century engineering, and to help readers begin along a promising path that we believe will lead to answers. The format of the discussion is *metaphor*, a primary tool of higher education. Metaphor provides a method to leverage current knowledge for the purpose of learning something new. Since the universe is full of systems, it was easy to provide many examples that show diversity as well as commonality among properties. From examples, we sought to discover abstract general concepts that led to axiomatic principles. To engage all thoughtful readers regardless of prior training, our approach remains largely storytelling and graphical throughout, using concepts already a part of the discourse on natural and manmade systems.

Chapter 1 provides background information using examples of systems already familiar to most of us. Here we hope to convince readers that Systems Engineering is a pursuit already widely shared by many fields of study, whether or not it is labeled as such. We frequently hear about persons attempting to engineer biological, agricultural, environmental, and social systems, but we hear little in the nontechnical literature about the general principles that all of these systems share. We aim to increase reader appreciation for the broad span of problems addressed under the label of complexity.

Chapter 2 discusses emergent properties, how all systems naturally seek equilibrium states, and how those states can slowly evolve and suddenly transition to produce different systems with new properties. Properties are the features that determine the response of a system to an input stimulus.

The discussion leads to four propositions on which we base our approach to system analysis and problem-solution development.

Chapter 3 reviews the methods commonly used to discover new knowledge, which are important to keep in mind when exploring unchartered territory. Here we introduce eigenanalysis as a tool for exploring system properties. We discuss the physics of scale and symmetry as fundamental for recognizing and categorizing properties. We also compare reductionist and non-reductionist approaches that science and engineering often adopt and contrast for such analyses. The strengths and limitations of both approaches can aid or limit progress toward discovery and understanding, and so each is to be applied carefully and appropriately.

In the second half of the book, Chapters 4-6, we get specific about modeling specific system types encountered in engineering analyses. Chapter 4 discusses transitions in mechanical systems, which are of great interest to traditional engineering education and practice, and which offer intuitive and concrete illustrations. We explain how mechanical systems are modeled as a means for understanding and predicting their responses to a stimulus. The utility of linear-system analysis is explained, despite the fact that only modeled systems can be truly linear. We also describe mathematical and computational modeling techniques that specify the eigenstates of a system and that combine to mediate observed properties. We show how complex systems are frequently simplified to generate linear models that provide insights into the inner workings, albeit a little biased, and how we must apply all models with care to avoid their limitations. Modeling the universe is how scientists and engineers learn about the systems that compose it. There is a dynamic tension pulling models toward simplicity to clarify the insights they offer at the cost of completeness and accuracy. Increasing model complexity improves predictability but often at the cost of intuitive clarity. The art of model building is a critically important skill for understanding system functions. Also, the fundamental roles of hierarchical system structures and internal variability are introduced for later exploration in Chapters 5 and 6.

Chapter 5 examines biological systems by applying the concepts of earlier chapters. The central idea is that the functional unit of life, the cell, is an information engine. A feature of cells that is unique among all natural complex systems is its robust drive to expand its information capacity. With these concepts in place, we briefly describe the field of systems biology as a re-envisioning of biological sciences as information systems responding and adapting to their environment. Systems theory is at the heart of the new science of molecular biology, and these ideas have strong implications for the future of medicine. We conjecture that new ideas emerging about the cell as a functional unit of life also extend to the role of people in organizations within society.

Chapter 6 attempts to bring all these ideas together in the context of the crucial role that variability plays in determining whether a system is simple, complex, or something in-between. Here we define the modeling concepts of system variables and parameters, intrinsic dimensionality giving rise to degrees of freedom, and how the types of variability within a system can lead to profoundly different emergent properties. We discuss how complex systems can have a great number of eigenstates, but if only a few states are accessible to the system for generating responses to stimuli, system behavior will be simple, predictable, and scale invariant. As more eigenstates become accessible to the system, in the sense that they express properties in the form of responses to stimuli, we find that system behavior becomes very rich, scale dependent and much less predictable – what we often think of as complex. These systems can still be monitored experimentally by applying a new type of eigenanalysis that we can only begin to specify today. Variability plays an existential role in the formation and behavior, and evolution of complex systems.

We tried our best to attach related ideas from the systems engineering literature, such as bifurcation theory, self-organized criticality and the law of requisite variety, to show a metaphorical consistency between our proposals and existing theory. In this way we hope to engage experienced scientists and engineers who may select this book as a way to begin a discussion with their students studying these and related concepts. We pushed the conceptual frontiers as much as we could without describing a mathematical theory to be sure to inspire discussion and, perhaps disagreement, and to not just report on the existing history of systems thinking. Many books already offer detailed historical perspectives on the development of systems thinking. We sincerely hope that all or part of the following discussion makes it into educational programs so that future engineering professionals are inspired to pursue the discovery of and applications of principles underlying all manmade and natural systems. Human society desperately needs engineers who understand the behavior of the complex systems that profoundly influence our lives and on which we all depend. Predicting, building, and managing systems is, after all, what engineers are trained do.

Michael Insana and Jamshid Ghaboussi, Urbana IL, September 2016

Contents

Chapter 1

Examples of Systems

1.1 What is a Complex System?

Have you ever been to a live college sporting event, like the basketball game picture in Figure 1.1? An event attended by thousands of enthusiastic fans? If so, then you experienced the great sensation of a crowd's thunderous cheer as the home team scores. Many who attend college games, in particular, are quite happy to endure inclement weather, traffic snarls, and high ticket prices to experience the comradery and the thrilling feeling of the roar of thousands of people rising to their feet and screaming at the top of their lungs. Crowd responses generate much excitement in each spectator, which further builds the intensity of the response. Even if you become distracted for a moment at a game, you are likely to find yourself compelled to respond just as excitedly to a play as those cheering around you because of some type of connection each of us feels to the crowd that surrounds us. Thunderous cheer is an emergent property of a complex system of spectators in a stadium that is responding to the action on the field.

We do not normally think of crowds of spectators as systems, complex or otherwise, perhaps because we don't often think of crowds as having unique properties. After all, these people mostly do not know each other. They each come to the game for their own reasons. Despite an apparent lack of central leadership or any agreed-upon organizational structure among the spectators other than shared decorum, the crowd naturally assembles as a group – a thing with distinct properties arising from the act of many individuals communicating in both subtle and overt ways. These self-organized interconnections enable the crowd to respond in unison to common input stimuli, and that response is an emergent property unique to self-assembled crowds. Crowd responses emerge from self-organized interactions

1

among individuals and are not a distinct property of individuals. There-fore, spectators are indeed a complex system. *A complex system is an assembly of interacting components each with variable individual properties that self-organize with little to no central control.* When the crowd-system receives input stimuli, it responds in a manner consistent with properties that emerge from its network of interacting fans.

Fig. 1.1 Cheering fans at a University of Illinois basketball game. Photo by Cary Frye/Illinois Athletics, December 2, 2015 (with permission).

Now consider a different complex system – a Midwestern Prairie. It too has properties not wholly due to any one of its components. The compo-nents of the Midwestern Prairie complex systems include various types of grasses, plants, animals, insects and micro-organisms. The interactions be-tween these components give rise to the emergent properties of the prairie systems that, among other things, determine how it responds to various climatic conditions such as extended periods of draught or heavy rain fall.

Grasses dominate life in a natural prairie, and they survive because of the support they receive from other plants and animals that live co-dependently. Each organism responds individually, cooperatively, and an-tagonistically with respect to others and to the environmental inputs. Prior to European settlement, prairielands in the Midwestern United States may

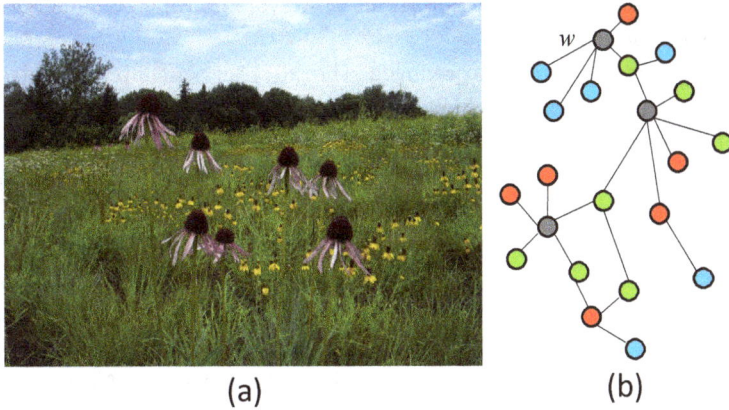

Fig. 1.2 (a) A patch of Midwestern Prairie in Illinois. (b) A "graph" of interacting prairie components is illustrated. Gray components are dominant grasses, green components are other plants, the red components are various insects, and blue components are environmental inputs like climate and soil conditions. Lines connecting component nodes, labeled w, indicate the strength of that interaction.

have extended as much as a million square miles (640 million acres). On the eastern side of the Midwestern prairie, including Illinois and Iowa, the prairie appeared to settlers mainly as a "sea of tall grasses," most 2-3 meters tall, consisting of mixtures including Big Bluestem and switch grasses that were mixed with a wide variety of wildflowers and a few trees. Today, that rich diversity of plants has been replaced by a crop grass – field corn – and its synergistic partner, soybeans. Only small isolated patches of the original prairies remain, although some local governments and citizen organizations are attempting to restore more of it.

Adjacent states to the west, Nebraska, Kansas, and Oklahoma, which were once dominated by a diverse mixture of medium-length grasses like Little Bluestem and wheat grasses, have now been replaced by mostly wheat crops. States in the western-most region of the prairielands, Montana, Wyoming, and Colorado, once populated by shorter Buffalo-grass varieties are now rangeland for cattle and sheep. The type of grasses that developed in a region depended on climate, soil type, and average moisture content. In each region, plants and animals were self-selected for their ability to live together and thrive despite climates of high wind, intense direct sunlight, and long periods of dry conditions. We might begin to analyze this complex network of components by diagramming the relationships among prairie species and their environmental factors using graphs like that in Figure 1.2.

The goal of the analysis is to determine how system properties emerge from component connections.

Human agriculture has reduced the diversity of life provided by prairielands considerably, which concerns environmental scientists. Diversity of species in the environment helps the prairie system ward off the spread of disease brought about by, for example, climate change and the introduction of non-indigenous invasive species. Component diversity and property variability are important features of any natural complex system. In Chapter 6, we describe how diversity and variability in component properties are essential for creating the system and maintaining a consistent "balance" so that system properties are able to emerge and establish a norm – an equilibrium state. Predicting the behavior of complex systems, especially those much more complex than human social behavior during a basketball game or even life in a Midwestern Prairieland, is at the heart of the most difficult problems facing society. For scholars in all fields, managing complex systems is the central intellectual challenge of this century.

Prairies are compelling illustrations of a complex system because they are familiar to those most concerned about the effects that civilization is having on our planetary environment. A prairie consists of and is sustained by its *biome*, which is a regional community of animals, insects, plants, and microbes that have, over time, both modified and adapted to the local soil, climate and terrain. Species both determine and are determined by the ecosystem they constitute. As stewards of our planet, mankind recognizes its obligation to identifying biome components and understands how they interact to balance competing forces within the ecosystem that sustain life on our planet.

1.2   Flock of Birds

The fascinating behaviors of flocking birds, swarming insects, and schooling fish are standard examples of complex adaptive systems. In each case, a countable number of components each follow a few simple individual rules to achieve complex group behavior. The simplicity of individual behavior provides insights into system properties and component interactions through modeling. And yet, this simplicity is the exception rather than a rule. Most complex systems, as we will see later, are too "complex" for us to clearly understand how their internal workings give rise to emergent properties.

Fig. 1.3   Two images of flocking birds. The top image is from Shutterstock. The lower photo is by Anthony Moyes.

Who has not marveled at the amazing aerobatic feats of hundreds, maybe thousands of birds as they seem to flow in lockstep through the air in intricate three-dimensional patterns? (See Figure 1.3). Birds use flocking behavior for many reasons, including protection against predators, efficient identification of food sources, minimizing the energy expended when flying long distances, finding a mate, and raising offspring. Depending on the task at hand and the environment in which the group finds itself, flocking birds execute aerial displays without coordinated leadership. Flocking behavior is a property of the system that emerges from individuals provided they are able to efficiently communicate and they adhere to a few simple rules. Deriving complex group responses from simple individual responses has fascinated scholars from all fields because most of us intuitively feel that this simple example could be a path to discovering fundamental principles governing all complex systems.

First consider a group of birds foraging for food. A good time to observe this behavior is in the summertime just after a large grassy area is mowed. If you casually walk by these birds that are busy eating newly exposed insects, you find a few skittish birds will fly away, prompting others to do the same. Others stay behind and ignore your passing. Of those that fly away, some may fly to treetops beyond your reach until you are gone, while others fly just a short distance and quickly return once you pass by. This variation in individual behavior can help systems remain in equilibrium despite a changing environment and, indeed, changes to the system itself.

Individuals in the group self-organize through visual and auditory signals to achieve synergistic goals, which in this example include locating food while monitoring the threat level. A property that emerges from flocking behavior is easier location of and safer access to food sources than might be expected by individuals venturing out on their own. It is significant that, although each bird has ostensibly the same instinctual rule set for how to react in the foraging situation, individuals freely respond with much variability to the same input. One can imagine that an appropriate degree of variability among the movement of individuals affords the group with greater opportunity for foraging success. Variation in behavior provides other advantages and challenges to the group.

The situation changes entirely when the group task suddenly switches from foraging to aerial evasion of a predatory bird. It is one thing for a group of foraging starlings to be wary of a slow-moving human walking by, and quite another when a hawk attacks them from the air. Once the hawk is detected, each starling narrows its list of possible rules of behavior to just a few so that variance among individuals is reduced. The hawk's strategy is to isolate an individual from the flock where it can use its size and speed to focus its effort. The starlings react instinctually to spatially vary their number density in the air while being careful to not separate from the group. They do this by continuously splitting and merging subgroups without separating from the flock. In this way they achieve their synergistic goal of confusing the hawk who the starlings hope will tire and disengage. What are the simple rules that enable instantaneous coordination of efforts even when the rules are intuitively understood and variably applied without the aid of central leadership?

This was the question asked by Andrea Cavagna, a condensed-matter physicist living in Rome. He leads StarFlag, a multinational collaboration of scientists dedicated to the study of bird flocking behavior because of what it can teach us about complex systems [Attanasi *et al.* (2014)]. The group's initial effort was to image in time the movements of each bird within a large flock. They hoped to understand flock shapes from individual bird movements using probability theory in a manner analogous to the way bulk material properties can be predicted from molecular movements in statistical mechanics. The StarFlag group inferred the rules of individual-bird behavior that could be synthesized in a computational model to study and predict responses of the flock to inputs similar to those observed in nature.

Bird movements were measured in three dimensions using *stereoscopy*. Each frame in a stereoscopic movie consists of a pair of images recorded simultaneously from perpendicular angles. These two views allowed the scientists to track the location and velocity of more than 1000 starlings in a flock under different environmental conditions. They observed that as a hawk approached, the starlings closest to the hawk took straightforward evasive measures. But it was the behaviors of those around the point of attack that generated defensive flocking behavior.

They found that individual birds followed three basic rules: (a) their strongest impulse was to move away from neighbor starlings in closest prox-imity to avoid collisions (strong short-range repulsion); (b) their weakest impulse was to move toward the bulk of the group to avoid separation (weak long-range attractions); and (c) in between the short and long ranges, star-lings moved in patterns most sensitive to the direction of the flight of the nearest 6-7 neighbors, preferring to fly closer to neighbors on either side of them than those in front or behind (intermediate-range alignment). Inter-estingly, the lateral intermediate-alignment rule correlates with the angular variation in the visual-field sensitivity of a starling. One can debate whether the high sensitivity for seeing birds along each side is a cause or an effect of the intermediate-alignment rule. By following these three simple rules within the physical limitations of perception, cognition, and flight, the flock was able to rapidly vary bird density and thus minimize risk during an at-tack. These rules are representative of the type used in agent-based models as discussed in Chapter 6.

It is also interesting how differently the flock behaves in the foraging and predator-attack environments. We would say that the onset of a hawk attack triggered a change in flocking behavior, where the same birds are now expressing very different emergent properties because they are behav-ing as part of a complex system. In the defensive state, individual behavior narrows to focus on three basic rules until the threat passes and the sys-tem returns to a more relaxed state. The effects of a tightened rule set combined with perception-reaction delays and flight maneuver times produce the rapidly varying bird density. This defense strategy must be robust against birds leaving and joining the flock. If the hawks adopt pack-hunting strategies, which may negate the primary effects of flocking behaviors, then starlings must also adapt.

The starling's defense strategy raises an important question: why do predators continue attacking bird systems that have evolved flocking behavior to protect themselves from predators? If the predators are not successful at achieving their goal of preying on members of these systems, it is reasonable to assume that they would eventually realize this and stop attacking these systems to avoid wasting energy. But they continue attacking them. Now we see another role for variability. Flocks of birds and schools of fish have evolved to provide protection from the predators for most, but not all, of their members. Some members become weak, old, or sick and so may not be able to keep up with others over the extended period of time needed to evade attack. They will separate as the flock tries to operate outside the capabilities of these individuals, and hence they are more likely to be lost to the hawk's attack. By continually attacking the flock, predators cull the weakest individuals and thus become a driving force in the long-term evolution of the starling species.

Bird flocks, fish schools, and insect swarms are relatively simple complex adaptive systems, which is why many authors select them to illustrate their basic features. The ability of a system to adapt to environmental stimuli, by the system slowly changing as a direct result of the stimulus, is a defining feature of life. But nature is filled with much more complicated and interesting systems.

1.3   Early History

As academics, we are occasionally asked about emerging ideas and their potential influence on society decades from now. It is thought we might know something about these trends because we spend so much time looking for *isomorphisms*, which are common features shared by disparate problems. As it will become clear throughout this book, we feel strongly that the development of new methods for studying and analyzing complex systems is among the most important problems of the 21st century, but this is not a new problem.

Beginning in the 1930s, biologist Ludwig von Bertalanffy began formulating the physical principles governing biological-system *ontogeny*.

Ontogeny is the study of how an organism develops from a fertilized egg through adult maturity, and at the time scientists were just beginning to organize details of a field now known as *developmental biology*. The ideas that emerged from his studies blossomed into what he called the *General System Theory* that he felt had very broad applications in nature [Bertalanffy (1950, 1951, 1972)]. To explain Bertalanffy's contributions, we first discuss a few statistical facts about developmental biology and some college-level physics.

It is incredible to think that each of us begins our development from just a single cell (zygote). The large, roughly 0.2-mm-long zygote weighs just 8 micrograms (0.000008 g). It voraciously consumes mass and energy to repeatedly divide over 9 months to result in a 3000 g infant composed of about 3 trillion cells (3,000,000,000,000). Not only do these cells quickly multiply, most cells precisely specialize their form and function, i.e., differentiate. Cells also move among other cells and orient their positions relative to an accumulation of like cells to form organs, limbs, and all anatomical features we recognize as human. After birth, the nascent body continues its growth and development, such that in two decades the 80 kg adult human will consist of roughly 80 trillion cells plus an additional 800 trillion bacteria, fungi, and archaea of more than 500 different types. These other organisms are so small that they add only about 2 liters to the body volume. The genetic material from the microbiome (our nonhuman passengers) augments our own inherited genes to contribute many essential bodily functions like drug metabolism, nutrient generation, vitamin synthesis, immune-system support, and to generally maintain the ecosystem within us. Thus our bodies are a synthesis of different organisms (only about 10% human by number!) that have evolved together to achieve a truly wondrous harmonious balance – a healthy equilibrium state we call *homeostasis*.

Bertalanffy studied cellular processes in search of the general principles that describe how organisms assemble from a collection of dividing cells. Specifically, he developed a set of thermodynamic equations for systems that describe how cells self-assemble to achieve multi-cellular group functions capable of maintaining a robust, dynamic equilibrium state. *Thermodynamics* describe the macroscopic physical processes occurring within a system that determine how energy flows among molecules to generate forces that form and erode structures.

The second law of thermodynamics was a little troubling for biologists. It tells us that an *isolated closed system*[1] that is not in thermal equilibrium[2] will tend, over time, toward a spatially-uniform equilibrium state. That is, heat in the system becomes uniform so there is no internal thermal structure. The second law of thermodynamics further tells us that if we use a heated system to do work, e.g., using a gasoline engine to turn the wheels of an automobile, there is always some thermal energy that cannot be converted into work. There are intrinsic inefficiencies because some of the energy is inaccessible for the purpose of work. The inaccessible energy is equal to the product of temperature and *entropy*.

It must be said that interpretations of the concept of entropy, as it affects daily life, are still hotly debated even if the underlying concept is well-defined and fundamental. Essentially, entropy tells us things about the *information* contained in a system. If you disconnect the cable to your television set, you stop seeing pictures and see only noise patterns. You have gone from a high-information and low-entropy condition with the cable connected, to a low-information and high-entropy condition with the cable disconnected. Entropy informs us about how much information is *missing* from a system. This explanation of information has a rigorous basis in probability theory that we won't discuss in detail here.

If we find there is a hot spot in a thermos of water, we know something about the location where the water was heated. At a later time, we are likely to find this isolated water system is now at the same temperature throughout, and those temperature measurements now tell us a little less about that system. The thermally homogeneous water system has higher entropy, more missing information, than the heterogeneous system with hot spots. (It can be hard to think about a quantity describing something that is missing!) We also can expect the former system that is thermally heterogeneous to be capable of doing more work because a greater fraction of the heat energy is accessible for work. This low-entropy system has more thermal structure and hence more information.

Analogously, consider a piece of excised muscle tissue. It has more internal structure (lower entropy) than the same volume of tissue hours after it was placed in an enzyme solution like a meat marinate until many of the proteins in the muscle have dissolved into an amino acid soup (higher entropy). Human bodies are highly structured assemblies of interacting

---

[1] a system that receives and outputs no mass or energy

[2] Imagine hot and cold spots within a perfectly insulating thermos of water giving the system of water molecules thermal structure.

molecules and hence are in a low-entropy state; they are complex adaptive systems of molecules that store enormous amounts of internal information.

Since the universe is the only natural closed system we know of, the second law of thermodynamics tells us that the universe will, over time, tend toward a smooth structure-less system of atoms with maximum missing information (maximum entropy). Ugh. Why then do we see living systems naturally build and develop structures without external guidance or prompting? It is the nature of all living systems to build structure by taking molecules in a high-entropy state from the environment to assemble intricate and specialized protein molecules that exist in the body at a low-entropy state (high information content). Bertalanffy wondered why all living systems behaved so differently from nonliving molecular systems.

The simple answer is that organisms are not closed systems. Cells use their DNA instruction set and their manufacturing machinery to assemble parts for new cells from raw materials that happen to float by. These raw materials (proteins, sugars, oxygen, etc.) are delivered to the cell by the blood stream. They are picked up for circulation as blood flows through the gut and lungs. The process uses lots of chemical energy to accomplish this assembly work at great expense to the environment. Organisms reduce entropy internally by constantly assembling and distributing molecules to facilitate the growth and development that establishes homeostasis. Organisms are entropy pumps in that environmental entropy increases as it is reduced in the organism. So there is a net increase in entropy, and the second law of thermodynamics is satisfied. This information was known to Bertalanffy as he began his investigations.

Bertalanffy admitted that science owes many of its successes to the *reductionist methods* of system analysis. Reductionisms is at the heart of how new science is discovered using the scientific method. In the reductionist approach, systems are broken down into elementary components that are studied separately. By systematically isolating each system component, the number of dependent influences on that component is minimized and it becomes easier to understand how a component functions. In this manner, we discover the mechanisms underlying component functions within a system. Then, the well-defined components are reassembled conceptually in models to explain properties of the whole system.

Bertalanffy noted that having a descriptive catalog of all the parts of a complex system is important for understanding mechanisms, but it does not ensure we have captured the essence of system functions, especially when one considers very large complex systems with dominant emergent properties, such as those of organisms and the societies they create. He concluded that the isomorphisms among the natural and social sciences strongly suggest there must be a common set of principles governing the behavior of all systems, which he labeled the *General System Theory*.

We can interpret Bertalanffy's concepts using the terminology about complex systems developed above. He noted the reductionist approach to characterizing organisms was limited in its ability to describe whole organisms. Organisms operate as *open complex systems*[3] in which *feedback* plays a major role in maintaining an equilibrium state. At the cellular level, feedback allows cells to sense their equilibrium state as well as their environment and then affect changes in the cell that maintain or even transform their state as necessary to preserve equilibrium. Reductionism has played an important role in the development of *systems biology* by helping scientists reveal the fundamental component mechanisms used by cells through their expansive networks, often labelled *gene circuits*. But that success has yet to fully translate to the scale of whole multicellular organisms.

At the organism level, feedback enables homeostasis as it does in the cell but it can do much more: it equips an organism with the ability to seek information, and that ability is a powerful tool that allows the organism to direct its course through life. Today we call a complex system with advanced feedback and memory mechanisms a *complex adaptive system*. Adaptivity is a property of organisms not always identified in the reductionist approach, but it is entirely consistent with the non-reductionist viewpoint where the adaptive property emerges from whole systems. Bertalanffy offers as evidence the fact that physiological and psychological stresses are not always something that organisms avoid, and yet one might expect stresses to be minimized by systems that strictly follow thermodynamic principles. He notes that sometimes organisms achieve a higher form of life through the use of feedback mechanisms that lead to increased stresses. He hints that this may be a driving force in evolution that cannot be explained using reductionist methods.

While many of these ideas were originally developed to explain ontogeny, Bertalanffy quickly recognized that the same concepts apply to system behaviors in psychology, neuroscience, and the social organizations. We too are compelled by the isomorphisms that appear among all complex systems, and we aim to contribute to a general theory by discussing common principles in later Chapters.

---

[3]those that freely exchange information in the form of energy and mass with their environment

First, we note that experts already routinely predict the behavior of complex systems using non-reductionist approaches, although not always reliably. They use *test statistics* measured from a very sparse sampling of system observables. For example, modern farmers weigh the measured conditions of the land available to them; as well as climate, labor, and market forecasts; and their tolerance for risk before they decide which crops to plant each year. Physicians are trained to talk with patients for a few minutes, take a family history, a few physiological measurements, and maybe a blood sample before assessing health during a physical exam.

The common goal in both situations is to predict the future behavior of very large complex systems that are each only partially understood based on combining a few key measurements with the accumulated experience of an expert. So it is possible to make generally good decisions about the future behavior of systems at affordable costs without having a complete understanding of the system, and such decisions are now routinely made within acceptable error rates. In fact, the expectation of any career professional is to "make the call" based on a very sparse sampling of system properties and lots of accumulated experience.

1.4 A Special Complex System

Life and all the many layers of organization that living populations create are the most complicated, interesting, and challenging complex systems we know. The study of an evolved multi-cellular organism and its relationship to other organisms in the environment is the domain of biological, environmental, social, and medical sciences. Multi-cellular organisms are open complex adaptive systems of cell systems; they are systems of systems. While each cell is itself a complex system – a complete factory for intra- and intercellular communications, specialized functions, and cell-parts manufacturing – cells need to communicate with each other if the organism is to survive. Cells communicate through vast networks of chemical, electrical and mechanical signals. These signals enable a cell to sense the needs of the whole organism so that it can "decide" how to specialize its form and function. Multicellular organisms are considered a higher form of life as compared with single-cell organisms, like bacteria, because the larger number of components generates a broader range of emergent properties that provide a far greater range of functions. Since cells found in different organisms tend to be about the same size, the range of emergent properties from an organism tends to scale with cell number and organizational complexity.

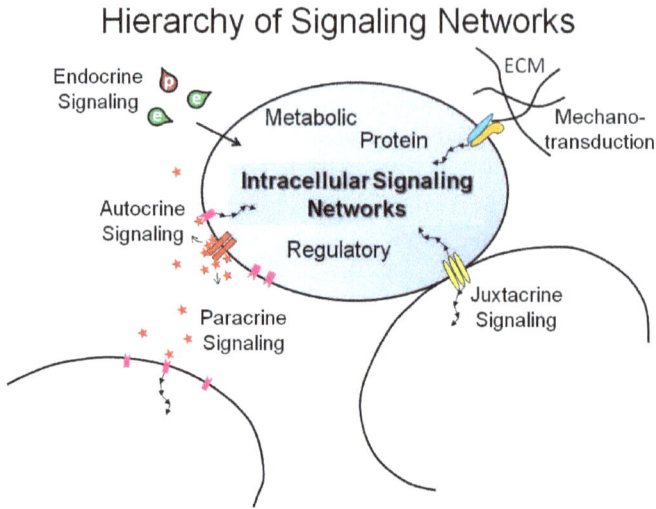

Fig. 1.4   Some human cell signaling networks are illustrated. Signaling occurs at many scales. Autocrine signaling is a process where a cell signals itself, and juxtacrine signaling occurs between cells in direct contact. Paracrine signaling occurs between nearby cells, and endocrine signaling occurs from signals secreted in the body centrally and distributed through the circulatory system. Cells can also be signaled mechanically by adhesion with the supporting extracellular matrix (ECM).

The complexity of the human organism lies in the rich symphony of multi-scale cell networks at play throughout the body (Fig. 1.4). Some networks move information among cells via mass and energy exchanges, others regulate manufacturing activities within a cell, and still others provide the energy required of all activities. Together they enable environmental inputs to a cell to combine with the basic cellular machinery and its unique DNA programming to define the activities and lifetime of each cell and ultimately the organism. Hence an organism is a complex system of cooperating cellular machines that self-organize according to the evolved rules of inherited genetic programming and the physics of environmental signaling. *Genes* are segments of the inherited DNA molecule that code the instructions for all cellular activity.

Cells are information engines that make life possible, and hence they are very special systems in nature. Each cell contains genes that hold a complete instruction set for building an entire multicellular organism. Genes in a cell code for what is possible within an organism, just like software in a computer determines what tasks a computer is capable of performing.

We can measure which cell functions are currently active in our body by measuring how genes in the cell nucleus interacts with the signals that can trigger those functions. Organ systems are an organism's interface between its cells and the ecosystem of its environment. In this context, the biological "environment" is shorthand for the whole hierarchy of complex systems that influence cell systems. Understanding the simultaneous interdependence of cellular machines at many spatiotemporal scales makes understanding the emergent property we call "life" so incredibly challenging and intellectually stimulating. We explore living systems in more detail in Chapter 5.

1.5   Summary

The examples above suggest that complex systems of all types have important features in common. Each system has numerous self-assembled components that are interconnected in a manner that enables them to transfer information often by exchanging mass-energy. The self-organized network of information that forms leads to emergent properties. When internal or external input stimuli are applied, a complex system responds in accordance with these properties. Emergent properties are responsible for the output behavior of a complex system that cannot be explained from properties of individual components; they arise from component interactions. Also the properties of a complex system are intrinsically related to its equilibrium state. System properties may change gradually as its components and their connections gradually change. However, that same system can also suddenly transition to a new equilibrium state depending on its stability.

Experience shows that variability among individual component properties and their interactions allow complex systems to maintain a robustly persistent output response despite fluctuations in the input stimuli from the environment. For example, organisms remain in homeostasis despite variations in available diets and feeding schedules because organisms have at their disposal a variety of energy storage and conversion mechanisms. In fact, variability, redundancy, and multi-functionality of complex adaptive system processes are hallmarks of successfully evolved organisms. These important aspects are explored further in Chapters 5 and 6.

1.6   A Broader View of Systems

Systems are the subject of intense scientific study because of how important they are in our daily lives and for what they tell us generally about nature.

Some natural and manmade collections of components seem too simple to garner much interest from the scientific community and are not generally considered to be complex systems. Let's explore some of these examples to decide if and when they can also be described as a system.

We claim that a piece of rock qualifies as a complex system. Its components are atoms that share orbital electrons to form a solid mass sometimes with fairly regular structure. Physics explains the ionic bonding processes whereby atoms share electrons, and thus self-assemble into an interconnected regular formation of atoms. Their connectivity yields emergent properties, like hardness, density, and melting point. For example, rocks are deformable. If we squeeze one a little, it deforms a little to spring back once released. If we repeat this process many times, we will eventually change the internal structure of the rock by adding tiny stress cracks, until one day we squeeze it just a little and the rock shatters, crumbling to gravel. If heated, our rock will melt at a temperature that depends uniquely on the details of its self-assembled structure. The emergent properties of rock are quite useful as road and building materials, and for extracting metals and precious gems. Although it may not be immediately obvious, we can say with confidence that even something as simple as a piece of rock qualifies as a complex system. If we grind the rock into dust and separate the particles so there are no longer significant interactions, then the disassembled rock may no longer constitute a system.

Now consider another form of dust, the most isolated collection of components we can think of: interstellar dust. As remnants of the big bang, vast clouds of dust particles can be found throughout the "empty space" between galaxies. This dust is composed of an assortment of atoms but most of it is hydrogen, the simplest element. Unless the dust particles are provided with outside energy, like when galaxies collide, the sparse clouds just float quietly in interstellar space, unseen at a temperature near absolute zero. Surely this dust cannot be a system, right? There are no significant interactions among these dust particles, they form no recognizable structures, and there is little available energy in the cloud to do any work. Since the dust cloud offers no information except atomic structure, the entropy of quiescent interstellar dust is quite high.

Then suddenly an old star in the galactic neighborhood goes supernova, and the ensuing massive explosion sends out waves of energy that eventually encounter our vast cold dust cloud. The ripples of energy passing though the cloud compress some of the dust particles together to form weak local gravity fields that draw in more of the surrounding dust. The accumulating ball of hydrogen dust continues to slowly grow until it is so big that pressure at its center ignites the hydrogen in nuclear fusion, and a star is born. Other clumps of dust might also accumulate around the new sun to form planets and moons, and millions of years later a solar system becomes clearly recognizable.

Even cold intergalactic dust has the potential to become a system. The energy waves it encounters initiate processes that lead to the formation of matter structures. These structures make the previously inaccessible internal energy of the dust cloud accessible so that work can be done. The work it does is to form a solar system according to the four fundamental forces of nature. If you study planetary physics, you will quickly find that many processes within the solar system, like tidal forces between planets and their moons, qualify solar systems as complex systems.

The solar-system example shows that the inaccessible energy in high-entropy collections of simple components with no discernible connections between them can become accessible to form vast dynamic systems with clockwork-like order and precision. Yet, solar systems age and die as their suns run out of fuel, and some disperse their atoms back into interstellar space in spectacular explosions to begin the cycle anew. Just like stars, living organisms are born, develop, die, and decay into dust, which appears to destroy the connections among components that formed the system it once was. Until that dust is picked up by an organism or solar system to be assimilated into its being, renewing the cycle.

Everything in the universe plays a part in the continuous lifecycles of complex systems. Once-isolated components can form connections to become system components. During its lifetime, a system can remain in a static equilibrium state, it might slowly evolve in dynamic equilibrium, and it can quickly transition to new equilibrium states; the fate of the system depends on its internal topology (the geometry of internal connections) and inputs received from an environment composed of other complex systems. The simplest complex systems, like rocks, can remain in a static equilibrium state requiring little or no energy to maintain that state for a very long time, even if they find themselves in a highly-variable environment. The most complicated systems, including humans and the organizations we make, require vast quantities of mass and energy to develop and sustain their very dynamic equilibrium states. Eventually all systems end their cycle and disperse their components. Cycles of complex systems are a fundamental structure of nature.

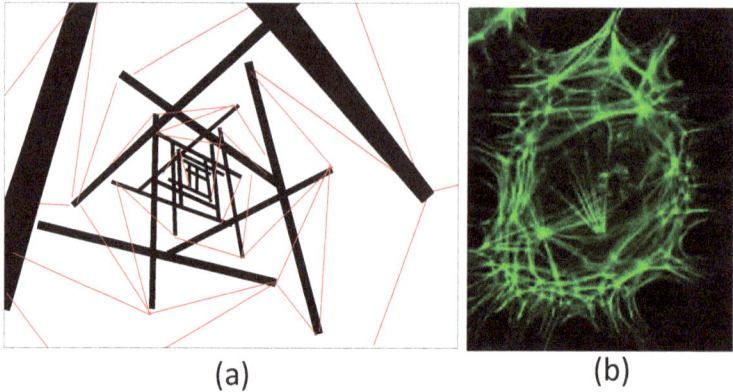

(a)   (b)

Fig. 1.5   (a) A drawing of a sculpture shown at the Hirshhorn Museum Sculpture Garden in Washington DC that was assembled using the principles of tensegrity. (b) The cytoskeleton of a human cell resembles the sculpture drawing in (a).

1.7   Types of Equilibrium States

Systems can exist in *static* or *dynamic* equilibrium states. A bicycle system parked in a garage is in static equilibrium because all of the forces influencing the ways its components can move are offset, resulting in no net motion. Of much greater interest is a bicycle-rider system moving at constant speed. This system is in dynamic equilibrium because the form and function of the whole remain relatively constant despite the rapid movement of its parts. A healthy Midwestern Prairie is in dynamic equilibrium when the *average* populations of its indigenous species remain relatively constant even though an insect species that is rapidly reproducing is just as rapidly being eaten. Many things, from the cells in our bodies to the stars in the sky, express properties that are a direct result of their dynamic states of equilibrium.

The concept of dynamic equilibrium is further illustrated by manmade structures assembled using the principles of *tensegrity* (tensional integrity), a term coined in the 1960s by Buckminster Fuller, who was an architect, inventor and systems theorist [Steven (1989)]. Special structural systems, like the sculpture drawn in Figure 1.5a, have solid components that never directly contact each other. Overall structural integrity derives from a balance of compressive forces applied to the solid components through tensile forces in the stretched cables. A desirable emergent property of this structure is its flexibility in the wind without overall loss of form. The balance of tensile and compressive forces in tensegrity structures is in contrast to conventional designs of structures, which are also complex systems with many connected interacting components but with different emergent properties. The solid elements in conventional building construction are connected directly to each other, and hence they are subjected to bending forces in addition to compressive and tensile forces.

Donald Ingber, a cell biologist and bioengineer, applied the tensegrity concept to explain functions in living cells [Ingber (2008)]. He describes how the cell's cytoskeleton Figure 1.5b, membrane proteins, and other specialized proteins work together with machine-like precision to provide the broad range of cellular structures, stiffness, movements, and other functions we observe. It is not just mechanical forces that must be in balance to maintain a healthy cell, but also the chemical and electrical forces that govern its behavior too. This rich, self-regulated orchestration of very complex cellular movements, signaling patterns, and parts manufacturing that cells in our body continuously conduct is absolutely essential if the complex system of the human body is to maintain homeostasis.

1.8 Complex-System Hierarchies: Systems of Systems

A modern example of a manmade complex system is a smart phone. The components within one handheld device are a collection of tiny electrical circuits that exchange information through the patterned flow of electrons acting in response to input signals received by the device. The internal communications within one device are managed by embedded software but its range of responses depends very much on signals originating from outside the device. Phones exist to send and receive radio waves and interpret responses with audio- and video-pattern outputs designed to be easily understood by humans. Each device is itself a complex system because its components respond synergistically to environmental inputs to produce outputs possible only through component interactions. Yet, if we really want to know how a smart phone works, it is not sufficient to limit our understanding to the operation of integrated circuits and software found in any one device.

A phone is an example of an *open system* because it needs information in the form of energy from its environment to function. Conversely, you can play video-games on smart phones that do not require connections to the communications network. In that situation, the phone is operating as a *closed system* that does not need mass or energy not already contained in the device.[4] So, a smart phone is an open complex system that functions as advertised only in the context of a larger complex system, the communications network. But why stop there? We have seen recently that the voice and data functions of smart phones placed in the hands of whole populations can drive economies and even contribute to the overthrow of governments. So when people say that smart phones are a *disruptive technology*, they mean that phone use by the world's population can couple large complex systems together in unpredictable ways, perhaps causing some of them to undergo transitions from one equilibrium state to another.

When we undertake a comprehensive analysis of one complex system, especially those found in nature, we find it difficult to know where the boundaries of the system are in relation to other complex systems. A detailed description of an open system requires one to carefully define its boundaries to specify which components are within the system and which components are part of the system environment that act as inputs to the system. The universe may be thought of as a closed complex system composed of a spatial hierarchy of open complex systems (like nested Russian dolls) that interact with each other. However, boundaries of individual systems are more to aid in human understanding than a description of physical attributes.

Since the definition of sub-system boundaries can be flexible, the components that we choose to include when analyzing complex system structures should change depending on the questions we ask. If we ask "How does a smart phone receive input signals and display them as output?" we might focus attention on individual devices to find answers. However, if we ask the more general question "How am I able to punch in a few numbers and direct a call anywhere in the world?" then we must consider a much larger system for the analysis that might include sub-systems segmented into service providers or countries of origin and destination. Indeed, if we ask the still-larger question "How has the use of cell phones influenced global commerce?" then we need to consider an even larger number of components where cellular communication devices and their networks might be combined to form but one of many components. Coarse-grained models, where many small-scale components are lumped into larger components based on their common function, are useful for studying large systems especially

---

[4]A true closed system is completely isolated from its environment. Playing a video game on a cell phone uses electrical energy that is dissipated as heat and radiated into the user and the environment. We are ignoring the transfer of all incidental electromagnetic radiation to make a general point.

when the fine-scaled details of the system are not relevant to questions being posed.

Specific answers about complex systems are obtained by asking specific questions. For example, if we ask the broad scientific question, "How does the extinction of one species affect all life on Earth?" we might not be able to provide a comprehensive answer because any imposed boundaries unfairly restrict answers to this broad question. If we instead ask "Can we predict how the transition of one cell to a cancerous state progresses to threaten the life of an organism?" we can find answers, and those answers have already led to new strategies for diagnostic and therapeutic approaches in medicine.

Truly-isolated closed systems are not found in nature, but they are very useful nonetheless as conceptual templates for developing analytical models that aid engineers, scientists, and medical professionals in their study of specific systems. Models of closed systems are formed by analyzing measurements of real systems acquired under various isolated conditions to identify individual components of the system and then map the relative weights of their interconnections. For example, species within a biome define the components of an ecosystem, their connections describe how they interact with each other (ecosystem topology), the inputs include the environmental factors like weather and soil, and the outputs might include the influence of that ecosystem on adjacent regions. Connection mapping among system components can be a daunting task for very large complex systems. The inaccuracies of system models are generally caused from the analysts' lack of knowledge of contributing components and their connections. It can be nearly impossible to determine the detailed topologies of some large systems.

Consider, for example, how difficult it must be to map the possible financial connections among the world's population of 7.4 billion to forecast long-term economic trends. We might narrow our models to focus on just one sector of the economy, thus sacrificing some generality of the predictions to achieve a manageable tool. We might also be uninterested in fine-scale details of economic activity, and settle for the aggregate activities of coarse-grained models. For example, instead of modeling individual people, we might measure the net economic activity of cities and use those groupings as components to study world economic trends.

Scientists are frequently faced with the problem of how to aggregate data for model building because those decisions influence model predictions. Imagine attempting to understand human brain activity by modeling the possible connections among the 90 billion neurons found in a volume of 1.4 liters that use the amount of energy consumed by a personal computer. This is the nature of the task assumed by neuroscientists. Even if an accurate and complete brain-network topology was available (it is not) and we had a computer large enough to hold all this data in memory, the neuron connections in a real brain are constantly in flux; they naturally form, strengthen, weaken, or disappear continuously as needed for brain function. The brain is but one organ in the body, which is just one organism in a population, etc. It is hard to know where the system boundaries are until one asks a question about this system. There are real limits to our ability to model some complex adaptive systems using approaches available to analysts today. Still, we go on trying because of our need to understand how brain connections generate some of the most fascinating complex-system behavior in the universe. Analysts are frequently tasked with predicting the behavior of large complex systems using partial mappings and even some incorrect information about network topology. Systems this large and complex, which are capable of truly amazing emergent properties, are mostly beyond scientific understanding using the analysis techniques of today. We need new analysis tools to gain deeper understanding of the most interesting complex system.

1.9   The Challenge of Complex-System Analysis

Dynamic equilibrium is the natural state of many complex systems. All living systems, and many manmade systems, require constant flows of mass and energy to and from the environment to maintain the steady-state responses characteristic of a system in dynamic equilibrium. In living systems, cells use environmental mass and energy to develop, build and maintain body structures. Information in the form of mass and energy is the currency of inter-cellular activities. Any well-designed manmade systems, and those in nature that have survived the rigors of evolution, will be robust to some variations in the flow of mass and energy on which these systems depend. However, systems occasionally transition to new equilibrium states that change some or all of the emergent properties. For example, consider the continuous need for balanced nutrition to maintain a healthy human, or the diverse blend of steady economic stimuli needed to sustain a healthy economy. Yet both systems can unexpectedly transition into unhealthy states. The challenge of complex-system analysis is to discover effective methods for predicting when transitions will occur and, if they do occur, then guiding the subsequent trajectory of system toward a new equilibrium state, preferably one with desirable properties.

Properties corresponding to a dynamic equilibrium state at one epoch of system history can be quite robust to relatively-large input fluctuations. These systems remain in equilibrium despite a highly variable environment. Over time, however, some systems can adapt to the variation by creating and connecting new components and strengthening or weakening existing component connections. These are evolutionary changes that modify system properties without the system transitioning to a new state. Then suddenly, the system might transition following relatively-minor input fluctuations very similar to those occurring earlier. What is it about the topology and dynamics of a system that triggers transitions at one time but not at other times? Are there early-warning indications that the system is nearing a tipping point? And, if reliable indicators can be found, can transitions be halted entirely or guided toward desirable states?

To approach these questions, let's consider the conditions leading to the worldwide recession of 2008. Recession is a period over which economic trade and industrial output of a nation decline. Inputs to the economy that triggered recession in 2008 were present decades earlier but the consequences then were less severe. For example, in the 1980's and 1990's we had an S&L crisis where 747 out of the 3,234 Savings and Loan Associations failed in the USA. There was a recession in 1990-1991 that some believe was triggered by this scandal, but that recession was much smaller and the economy quickly recovered. The difference in 2008 was the US economy had evolved to accumulate "fundamental weaknesses" as it adapted to the global environment at the time. The responses of the economic system to inputs led to economic-system modifications, e.g., weakening banks and changing government policies, to a point where the economy became vulnerable to a seemingly harmless trigger that transitioned it suddenly to a new equilibrium state with very different properties that the middle class continues to feel many years later. Similarly, the human body accumulates genetic errors with age that weaken it, making it vulnerable to environmental insults it would normally shrug off in youth. If we could identify the accumulation of "weaknesses" in either system we could predict the vulnerabilities to sudden transitions. We would then be closer to answering the big questions on the minds of all physicians, economists, and other systems managers today.

The roles of government policy and common financial practices as contributors to fundamental weaknesses in the economy are still hotly debated.

Did an amplified housing-market "bubble" encouraged by unwise lending practices create overall economic vulnerabilities? Did the debt-laden consumer economy that was driving the US markets for decades mean a recession was inevitable? Are growing income inequalities and a weak job market causes or effects of economic transitions? Is there a fundamental economic design flaw in our system caused by conflicts between rules of capitalism and the limits of democratic rule? Is the US economy too big to self-regulate, as complex systems naturally do, and therefore should we limit the occurrence of reckless practices with more government regulation? These questions have been debated for years, and believable cases can be made for conflicting answers. The uncertainty of how to prevent another economic recession has generated fear that has strongly polarized public opinion and crippled even the most basic of political functions in the US – budget formation.

We won't have a systematic approach to answering these important questions until we can find indicators of instability based on the general principles of complex systems. Experts do agree that the economic instabilities of 2007 and before that led to the 2008 recession were systemic, which means there were distributed weaknesses from long-term practices following some number of earlier events or policy decisions. Economic models aim to mimic the evolution of an economic system to anticipate the effects of singular and combined decisions on long-term stability. The collapse of the investment bank Lehman Brothers may have triggered the transition of the economy into severe recession in September 2008, but it was no more the cause of the recession than Rodney King caused the Los Angeles riots in 1992. Evolution of the US economic system had slowly changed its dynamic equilibrium state to poise it at the brink; the system had reached a tipping point where if one unfortunate event had not triggered the recession something else soon would have. And, since the world economies are coupled to some extent, all were vulnerable to the spread of recession.

The search for complex-system indicators has been going on for decades. Weather forecasts and Earthquake predictions rely on atmospheric and geophysical measurement as indicators of natural disasters. Medical research is, in part, a search for biomarkers as disease-specific indicators of patient health. Of course, economic indicators and medical biomarkers are discussed every week in the press. In fact, the search for indicators that predict the behavior of all manner of complex systems are on the minds of scholars of all persuasions.

Indicators are often statistical summaries of one or several combined measurements of system observables, which we call *test statistics*. In economics, we have the gross domestic product (GDP), the money supply, purchasing managers index (PMI), stock market indices, the consumer price index, and University of Michigan Consumer Sentiment Index as some of the most common performance-indicating test statistics. Changes in test statistics over time that fall outside of a specific range deemed "normal" are used to forecast changes in economic growth. Economic growth is the long-term ability of an economy to produce goods and services in a manner that raises the standard of living for its population. Imagine how many facets of society influence the US economic system's output!

The descriptive interpretation of a test statistic is tied to theories on the business cycle that yield models for predicting economic performance in which test statistics are parameters. If models are large numerical machines that predict the future, parameters are the knobs we turn to tell the model what we see happening today. Models can forecast trouble ahead and guide solution strategies provided the economy does not venture too far from established past behaviors on which the model is built. Models of complex systems that successfully predict behaviors in equilibrium states, often fail to predict the onset of sudden transitions or the emergent properties of the post-transition equilibrium state. This has caused the general public to be skeptical of model predictions by experts. In fairness, forecasting is an incredibly challenging task.

As we saw in 2008, standard indicators did not clearly forecast a major transition into recession or the stubbornly-high unemployment rate that was a prominent feature five years post-recession. In fairness, a few economists did predict the approaching recession. Scholarship is full of examples where mathematical and computational models work well in situations where the answers are already known except for a few details, but fail miserably at predicting a looming disaster of far greater consequence. If it is true that all complex systems share a common set of properties, then there must be a general method for finding indicators of equilibrium states that predict the occurrence and properties of sudden transitions. Going through any list of grand challenges facing society today put forward by institutions of higher learning, it is hard to find even one challenge that would not be advanced significantly by a breakthrough in the analysis of complex systems.

We believe that common principles governing all complex systems exist, and these principles point to a general approach to finding indicators. We describe these in more detail in Chapter 2.

1.10   System Response to Local Perturbations

One of the most important emergent properties shared by all complex systems is that the entire system can contribute to the response even when the stimulus is highly localized. When a change is introduced to just one region, a local response can usually be expected. Yet, over time, we often find that a much larger system responds, often with undesirable consequences. We have seen examples of this effect when policy makers attempt to solve a specific problem, like introducing affordable public housing in urban areas of the US during the 1960's. Initially, the effort stabilizes disadvantaged families in the region by providing quality housing for their families. However, housing is only one part of a very complex societal problem. When other critical elements like quality education, sustainable job creation, and neighborhood security are not also effectively addressed, the quality of life in public housing quickly deteriorates to the point where its presence becomes a disadvantage to the families and the community.

History is also filled with examples of persons who felt they understood ecosystems well enough to make "small" adjustments for the betterment of all. For example, in 1876, Americans brought the Kudzu vine from southern Japan to the southern US to help stem the effects of soil erosion. What these well-intentioned designers failed to realize was how competitive the Kudzu vine would be with their new regional plant species, and how well it thrives in the climate of the Southern United States. Every plant is able to grow about a foot each day, so this species is able to rapidly spread across a continent always at the expense of other plants unable to compete. The consequence is reducing diversity. The original intention was to deal with a specific problem but the solution selected ignored the potential effects on the larger complex ecosystem. The new plant interacted with other components in the ecosystem with unforeseen consequences, where gradually disadvantages overwhelmed any advantages. Generally, perturbing even a small part of a large complex system requires one to consider the response of the whole system.

In another example, twenty-four European rabbits were brought to Australia in 1859 by an English farmer hoping to give his Australian ranch a "touch of home." The rabbits did indeed feel very much at home; they interbred with native rabbit populations so successfully that their populations grew exponentially for decades when as many as 2 million rabbits could be eliminated annually without affecting the population. This example illustrates that a small input to a complex system – the seemingly trivial introduction of just two dozen rabbits to one farm – eventually resulted in a very large system response – explosive population growth resulting in substantial soil erosion and species extinctions across an entire continent in just 50 years.

In the 1970s, Americans introduced Asian carp into southern fish farms and sewage treatment plants to filter the water. These fish were a major success at their intended job, but a few escaped into surrounding ponds and eventually entered the network of American rivers where they eat up to 20 pounds of plankton and vegetation every day. They grow as large as 4 feet long and weigh up to 100 pounds. Asian carp (also called flying fish) pose a major threat to the Great Lakes if they enter those fresh waters. The river experience has shown that these fish will strip the lakes of the plant life that supports all fish. Again, we find that introducing just a few fish (small input) with no natural predators other than humans into an enormous complex system of continental waterways can begin well but eventually evolve the system to have undesirable properties. The effects in this case are to reduce plant and fish diversity.

These are just a few *simple* examples of the consequences of humans interacting with complex systems, intentionally and otherwise, that have resulted in major unintended consequences. Every day we discover how extensive the tether connecting all ecosystems extends. During 2013 the Detroit Free Press reported the discovery that pharmaceuticals, caffeine, artificial sweeteners, micro-plastic spheres, and even toothpaste that make it through water treatment systems are having a disturbingly profound effect on the Great Lakes. Even though the lake has 2,000 trillion gallons of water, trace elements of these substances are having hormone-like effects on fish populations that diminish their numbers. It is becoming clearer every day that every ecosystem, every species, every drop of terrestrial water shares a vital connection.

Yet, too often we fail to account for the "whole system". Instead we focus on smaller portions of greatest immediate concern. These are cases where reductionist approaches are insufficient to allow us to discover effective solutions. The affected systems are too large – often whole continents or populations – for us to have a complete mapping of the response. Local adjustments made with the best intentions can lead to larger systemic problems.

Today, the world's population is gravely concerned about the accelerating use of hydrocarbon fuels and its effect on greenhouse gasses in the atmosphere. These are the same fuels that generated the power that gave birth

to the industrial revolution and a century of unprecedented global pros-
perity. Civilization may have already released enough greenhouse gasses
over the last century to saturate the natural system buffers provided by
the atmosphere and oceans. We are concerned what climatic indicators are
telling us about the potential for irreversible transitions in global climate
toward undesirable states, just as we are concerned that rapid changes in
the availability of energy sources could generate a sudden economic down-
turn. Is human influence on the chemistry of Earth's atmosphere placing
us in peril of future sudden and radical changes in climate? While climate
change is almost certainly occurring, we do not today have the analytical tools to accurately pinpoint critical weaknesses in this large complex
system or to design effective and affordable interventions that won't also
generate unforeseen consequences. Our history of environmental tinkering
is just too checkered for the public to trust experiments in geoengineering.
Uncertainty breeds a collective anxiety that fans dissent between environ-
mentalists and industry, which helps no one. Both sides hope policymakers
can convince governments to make wise decisions about future energy use.
New analysis tools are desperately needed to guide policymakers.

Humanity's intense interest in complex-system transitions stems from
the catastrophes that can occur when we ignore the consequences of our
current practices with respect to economic policy, energy use, environmental damage from industry, and economic marginalization of large sectors of
the human population. Thoughtful people believe we can learn to develop
planet-management skills from past disasters, and there have been many.
For example, deep plowing techniques that were used prior to 1930 to raise
crops in the US were modified because of the resulting devastation wrought
during the dustbowl years in Oklahoma and Texas. Complex systems of
all types will naturally find a "balance" among its many component interactions, i.e., seek a dynamic equilibrium state, whether we intervene or
not. This is an important point because the *equilibrium state* gives a complex system its emergent properties. The tricky parts of managing a complex system are designing strategies that maintain the current equilibrium
state or guiding a transitioning system toward a new state with desirable
properties.

1.11   Self-organization Versus Central Control

We have seen that complex-system properties are a consequence of its self-
organized topology and the self-regulated interactions occurring among its

components. Self-regulation is a necessary element of all complex systems; conversely, collections of components that are entirely centrally controlled are not complex systems. We think of them as *simple systems*. The emergent properties of complex systems offer exciting possibilities, but they can also make it difficult to predict and control system behavior. Consequently, practical manmade systems introduce elements of central control to increase reliable predictability of system properties over time. Most human-designed systems combine elements of central control and self-organization depending on how predictable and innovatively adaptive the system needs to be to operate effectively in its environment.

For example, railway systems are predominantly centrally controlled. Necessity dictates that all train movements are scheduled and controlled centrally by a decision-making entity placed outside the bounds of train operation. This outside entity monitors and controls system activities to minimize human errors and mechanical failures within the system. When there are train accidents, one of the first questions asked is whether the cause was related to logic errors in the central control or to mechanical failure or "human error" in the small-but-influential element of self-regulation. Railway systems need to be as predictable as possible to be safe and effective. Minimizing variability in departure and arrival times is the central objective.

The vehicular highway system still has significant elements of central control (traffic lights, pavement structures, police enforcement), but compared to railways there is a greater requirement for self-regulation. Drivers play a larger role in directing car movements. Vehicles, drivers, and roadways are the most prominent components of the system. Traffic congestion and accidents are often a result of too many vehicles moving too quickly and independently for road conditions to be safely guided by self-regulated human drivers. The commercial interest in automatic braking-system technologies and the early trials of autonomous vehicles show a trend away from self-regulated human driver control of highway systems and toward greater dependence on sensors and driverless control technologies that add more centralized control. Commuter driving requires more individualistic scheduling than train travel, and yet major disadvantages appear when the highway system behaves with high variability and unpredictability. Traffic planners seek to optimize the necessary balance between self-regulation with central control in order to achieve the goal of safe and speedy transportation.

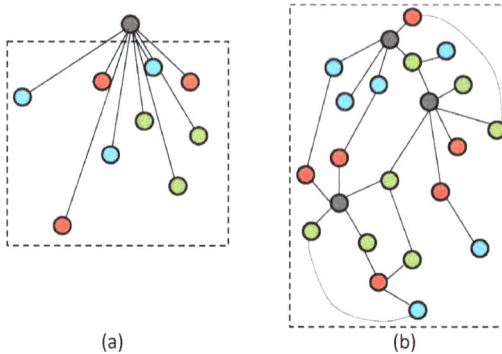

(a)   (b)

Fig. 1.6   Diagramming extremes. The components of (a) form a simple system where a centrally-controlled element outside the system influences each component in the system. The components of (b) form a system self-regulated elements.

Now consider the operation of a large publicly-owned corporation. Its employees and facilities are among the most important components of this system. Organizational charts often show a corporate structure led by a Board of Directors and CEO that ultimately are responsible for all major decisions. Large-scale central control is most visible to shareholders and the public. Yet, if we look more closely, we find that central guidance must cede control of day-to-day operations to departments if they are to be effective. Departments may be further broken down into working groups and individuals, where self-regulation becomes increasingly important at the finer scales of organization where the need for innovation and efficiency increases. There are major benefits to a corporation where individuals are empowered to seek the most innovative solutions to specific problems associated with product manufacturing. Subgroups self-assemble and dissipate as required to solve problems, and yet implementation of solutions must be guided by central considerations to ensure they meet overall corporate needs. We see that corporate central control in the form of vision and leadership is a guiding framework under which self-regulated components work so that the best overall solutions emerge. Unlike in the transportation examples, innovation is more critical to corporate success. Too much and too little central control inhibits overall corporation performance. The best corporations are able to strike just the right balance.

Universities, like corporation, have both control elements, although central direction is generally much weaker at universities than in corporations because of the greater need for creativity and less need for coherence of

thought. Young professors in engineering fields are hired with the expectation they will eventually self-fund and manage their own research laboratory
using university facilities. In a sense, they are asked to propose and solve
problems of interest to society through collaborations with industry, federal agencies, private foundations, and other universities. Their goal is to
produce and disseminate scholarship (new knowledge and devices) and a
skilled workforce, and the best universities give their professors maximum
freedom to achieve these objectives. A professor's achievements in research
are expected to eventually translate into the courses developed and offered
by the home unit, the mentorship provided to students and peers, and new
knowledge that s/he disseminates through many forms. Given the high
levels of innovation expected, university administration assigns day-to-day
control of operations to individual professors provided they are successful according to criteria established by the university. The self-organized
structure of universities is absolutely necessary to maximize its goals for
innovation and scholarship. Administration focuses its efforts on guiding
the overall direction of the research and the educational enterprises by selecting, promoting, and supporting with resources the most promising and
skilled professors and investors, which then attract the best students. Except for issues related to class scheduling, student registration, and facilities
management, predictability of university work product is valued much less
than the innovation born of independent thinking.

In these four examples of manmade complex systems, we demonstrated
that the needs for system predictability and innovation determine the appropriate levels of central control and self-regulation that should be infused.
Self-regulation in organizations leads to innovation because the system has
emergent properties, whereas central control leads to high predictability
because there are fewer emergent properties. Too much central control sti-
fles innovation just as too much self-regulation leads to unpredictable, even
chaotic, system behavior. Organizations of all types constantly debate how
to blend these two system-guiding forces to achieve their goals.

In this regard, socioeconomic systems are very interesting. Wars have
been fought and empires built and destroyed from humans experimenting
with these theories. Opinion ranges anywhere from free, self-regulated systems with minimal central control to those who feel there is an essential
role for a strong central government regulating everything from business
to personal behavior. Most acknowledge the need for some stabilizing
central control that maintains the dynamic equilibrium states of society.
Others worry about giving up self-determination that will stifle success by

permitting too much taxation and regulation – prices paid for central control. Too much or too little central control produces an unstable society, as we observe after years of autocratic rule or following a natural disaster in situations where law enforcement becomes dysfunctional for a time. The question for reasonable people is where to set various limits. These debates will continue as long as there are thinking people living in a free society. We believe that the principles of systems operation can contribute to these and many other discussions of great interest to all.

Chapter 2

System Properties

2.1 What Do We Mean by Properties?

Many examples of complex systems were described in Chapter 1. These and all systems express characteristic properties that emerge from the interactions among its many components. This chapter discusses the nature of those properties using examples from everyday life. It also describes methods for discovering properties of these systems by monitoring their response to input stimuli over time.

To begin, consider just one of many mechanical properties that can be attributed to a straightforward complex system – a wooden rod. If we use our hands to apply a force, as in Figure 2.1, the rod will bend a certain amount. Comparing two rods of the same length and made from the same wood, experience reveals that the rod with the larger cross-sectional thickness will bend less given that the same force is applied. The bending response of the rod system to the input force is a result of the rod's bending property called pliability. Pliability depends in part on properties of the individual plant cells making up the rod, the shape and orientation of these cells, how well they bond to each other, and the overall shape of the rod relative to the applied force. That is, overall system properties depend on the material and geometric properties of its components and their interactions with one another as they respond to stimuli.

Bending is just one of many mechanical properties that can be attributed to the rod. Knowledge of these properties helps furniture designers to use wooden rods to build larger systems, like dining room chairs. Additional knowledge of material properties of the wood helps these designers protect their product from environmental factors. For example, if the wood surface is sealed with paint, it can resist material degradation due to heavy use, humidity, and exposure to sunlight, insects and extreme temperatures.

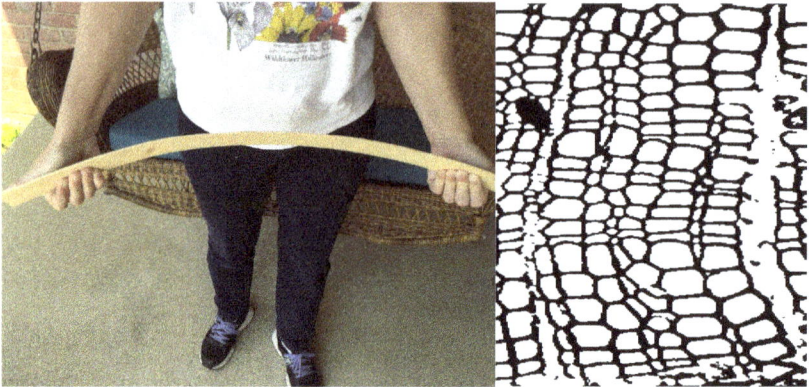

Fig. 2.1   A wooden rod is demonstrating its pliability property (left). On a smaller scale (right), we see the material is composed of bonded pinewood cells.

A wooden rod is a complex system according to the definition given in Chapter 1. Physical properties of its wood-cell components, the strength of their bonding interactions, and overall shape determine its mechanical properties, just as properties of the interacting rods in a chair determine the overall properties of the chair system.

Large or fast deformations can perturb the internal cellular structure enough to slightly modify its mechanical properties; the rate at which a property changes depends on the response of that system to all environmental inputs, not just applied force.

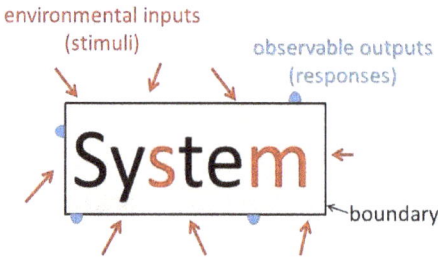

Fig. 2.2   Systems respond to and are changed by environmental inputs.

We use terms like "input" and "stimulus" to describe how matter and energy from outside system boundaries – from the environment – influence the system. Terms like "output" and "response" describe the effects of that influence (Fig. 2.2). Input stimuli interact with system properties to generate the responses we observe. Over time, perturbations to system structure can generate changes in some of the properties: weather exposure can dry, crack, swell, and rot the wood in ways that may not be immediately obvious until one day a routinely applied load causes the rod to fail.

The relevant properties of a wooden rod depend on the context in which they are being considered; among these properties are chemical, economic, and aesthetic properties, where the latter two are bestowed when humans interact with the rod as part of a larger business system. We cannot know of these system properties unless we apply an appropriate stimulus and observe the responses that are a consequence of that property. Experiments involve a sequential application of input stimuli to a system under observation that invoke the responses needed to discover properties or to confirm properties predicted by a model. Whether we observe a property or not, that property exists, which is important to remember when analyzing overall system behavior.

Before the wooden rod became part of the chair, it was part of a living tree system. As a tree part, the material had very different mechanical properties. Among these are greater mechanical flexibility and the possibility of growing, branching, and producing leaves that help facilitate the biochemical machinery of plant life.

As part of a living tree system, our collection of plant-cell components achieved an equilibrium state that changed slowly over years in response to its environment and internal DNA programming. At some point, the tree system transitioned suddenly to a new state as it yielded to a saw and a lathe to become a component of a chair system. The transition is considered sudden because it occurred during a short time period compared to the time the wooden rod spent as part of a tree. Now, as a component of a well-maintained chair system, the wood cells again slowly change over many years in response to its environment until it fails and transitions again ultimately, perhaps, to the state of firewood ash.

This long story about a wooden rod is offered as a template for analyzing larger complex systems. A non-reductionist analysis requires all systems to have properties, whether the systems are objects, plants, animals, ecosystems, or societal organizations. We routinely apply stimuli to systems that we encounter in our lives to measure their responses and thereby discover system properties. This process is followed for all manner of systems from wooden rods to political parties. For example, if a child appears sick, you might apply a hand to the forehead and then a thermometer under the tongue to sense body temperature with increasing accuracy before deciding if the child system is in a fever state that requires medical attention. Or, if you think you are ready to purchase a house, you might contact a mortgage broker and a real estate agent to inquire about the indicators of current market states. When these are compared with your financial state, you are able to make an informed decision about the timing of a purchase. Insofar as these stimulus-response experiments accurately represent the state of a system, we obtain valuable information about properties that can guide future behavior. The shortcomings of this approach become apparent when selected indicators fail to accurately define properties of the current equilibrium state of the system. In the remainder of this chapter, we look more closely at key attributes of the systems above.

2.2   Emergent System Properties

All of us use stimulus-response methods to discover the emergent properties that define systems. This approach works when experts study large complex adaptive systems, such as brains, flocks of birds, and economies, and it works when each of us examine simple everyday systems, such as our house, its furniture, and wooden rods within the furniture. All systems share fundamentals characteristics, such as being composed of many interacting components whose organizational structure generates the emergent system properties we observe. It is important to probe system properties at many spatial scales to understand them completely, from whole systems within their natural environments to isolated components of the system. Analyzing systems is at the core of engineering, just as knowledge discovery is at the heart of science.

We noted that system components and their interactions can be changed by internal perturbations, like component failures, or external perturbations like environmental stimuli. An accumulation of these small changes over time results in the evolution of emergent system properties. For example, medical practice recommends that we periodically introduce dead or weakened infectious agents into our bodies for the purpose of modifying our immune responses to fight off a more robust future infection. Also we attend school daily for years to be subjected to environmental stimuli that direct the formation of brain-cell connections and provide us with the stored knowledge and self-discipline required for mature decision making. Immunizations and learning are examples where stimuli applied to complex adaptive systems make small immediate changes that can develop to have profound influences on individuals over time, and, perhaps more importantly, they can propagate beyond the individuals to strongly influence the emergent properties of society.

Objects too can be affected by their history of environmental stimuli. For example, consider the industrial process of forging that is used to strengthen metals through cyclic applications of heat and deformation. While some materials, like wood, accumulate damage with repeated deformations, others become stronger. In both cases, the emergent properties of these systems can be found to evolve slowly over time.

Sometimes incremental changes to system properties can survive beyond the individual system receiving and responding to the stimulus. For example, strong immune responses and effective bird-flocking behaviors can be passed to offspring genetically even if the process requires many generations of repeated input exposure to express. Biological evolution tends to increase the likelihood of survival of a species. It is a well-known example of the slow steady progression of emerging properties.

However, there is a long history of man-made complex systems evolving over many life cycles. Consider the evolution of manufactured products where market pressures motivate manufacturers to strive for greater value at lower cost. Another example is the rapid growth and evolution of the Internet, where in just a few decades the complexity and richness of its emergent properties have exceeded everyone's expectations. The Internet is a global system that now touches all corners of human thought.

In general, *emergent properties* specify how a system responds to internal and external stimuli, including any perturbations within the system topology. Two important emergent properties shared by all systems may lead us to a new way of looking at systems.

The first important emergent property is that all systems are found in *equilibrium states*. Some readers may have some difficulty in accepting this as an important property because it sounds obvious — of course everything around us is in a state of equilibrium. What is special is that we exist with the properties we observe precisely because we and everything around us are capable of being in an equilibrium state.

The second important property also shared by all systems is their ability to undergo *sudden transitions* from one equilibrium state with one set of properties to another state with possibly very different properties. For example, bending the wooden rod beyond its yield point irreversibly transitions the rod into a state with much weaker mechanical properties. Sudden transitions are in contrast to the steady evolutionary changes in properties that any system in dynamic equilibrium can undergo. Sudden is a relative term, as we discovered when considering the transition of plant cells from tree to wooden rod. Sudden transitions move systems from one equilibrium state to another quickly, whereas changes brought about through perturbations occur within a system that remains in equilibrium, even if it is an evolving state. The accumulation of small system changes can slowly drift properties of the equilibrium state, even significantly, although the system remains in a state of equilibrium. A system is in a state of equilibrium when small structural changes or input stimuli generate small properties changes or output responses. What qualifies as "small" or "large" is often a matter of debate. Of greatest concern are those progressive changes in system properties that erode the stability of the equilibrium state to a point where a system becomes more susceptible to environmental triggers. Repeated flexure of a wooden rod can generate micro-cracks that accumulate until even a modest force induces mechanical failure. Also, the current world economic events discussed in the last chapter are examples of a drifting and destabilizing equilibrium state that can suddenly transition.

What are the changes that occur in system structure and properties that trigger a sudden transition? This question is of primary importance in managing organizations, economies, ecosystems, health, and other important systems. At the present time, our knowledge of systems is not comprehensive enough for us to answer this question in detail. Nevertheless, being able to predict sudden transitions is of primary importance in the study of complex systems. We describe a framework for addressing this question at the end of the chapter.

2.3   Systems in Equilibrium

The universe is composed of systems in various equilibrium states. Each system responds to input stimuli to generate responses according to its properties that emerge from its assembly of networked components that may each have distinct properties. System properties are characteristic of its equilibrium state, but that state of equilibrium can slowly change over time. System networks undergo structural perturbations from internal changes or in response to environmental stimuli. These perturbations generate changes in the equilibrium state that we see as variability in responses to a fixed stimulus.

Definition: *A system is in a state of equilibrium when small internal and/or external perturbations result in small changes in its properties and responses.*

For example, lifting weights regularly for years will prompt the healthy human body to add muscle mass, which is a progressive change in the healthy equilibrium state of muscle to repeated stimuli. The weight-lifting stimulus prompts internal changes in organisms that remain in homeostasis even as they exhibit an evolving property we call increased strength. Chronic stimuli induce somatic changes (affecting the individual) and potentially genetic changes (affecting progeny).

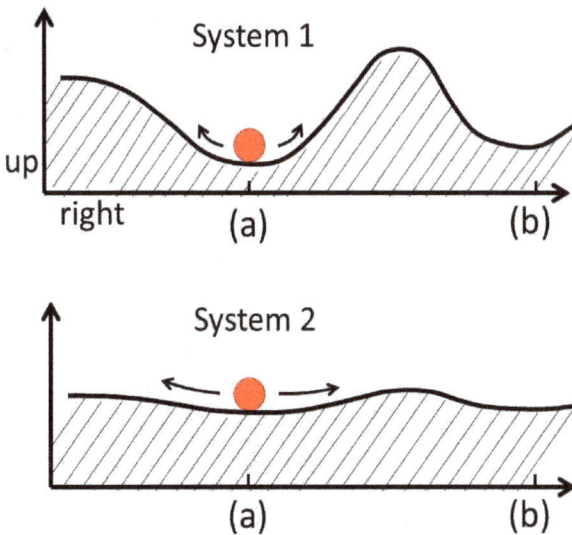

Fig. 2.3   Two systems in static equilibrium are illustrated. State (a) in System 1 is more stable than State (a) in System 2 because it requires more energy for the ball in System 1 to climb the gravity well and settle in State (b) than it does for the ball in System 2.

We monitor the stability of an equilibrium state by detecting changes in emergent properties through the responses of a system to stimuli carefully selected to probe that specific property.

Systems are constantly subjected to small perturbations internally or externally from the environment that may generate reversible or irreversible changes in system structure. The term *small* is in regard to the scale of the system as described below. *Reversible changes* may or may not cause changes in emergent system properties. However, *irreversible changes* definitely cause changes in emergent properties.

For example, Figure 2.3 shows a rolling ball in two different mechanical systems initially placed in a gravity-well equilibrium state (a). Balls can only move in the plane, left-right and up-down. We can say the ball in System 1 is in a more stable equilibrium state than the ball in System 2. To see this, imagine these simple systems are riding in the back of a truck that gently jostles both balls in their planes of motion about the equilibrium point, which is the bottom of the well.

The properties of these two systems are different. We know this because we observe the ball in System 2 rolls farther left and right than the ball in System 1 for the same applied forces. We also find in System 1 that the jostling causes the ball to roll right and left but it often returns to equilibrium state (a). However, with the same jostling in System 2, we are more likely to find the ball suddenly transitioning from equilibrium state (a) to state (b). The transition is reversible because the ball can return to (a), and when it does the rolling properties are the same. System 2 might be System 1 viewed at a later time, once it has changed to develop properties with greater instabilities in the form of lower positional transition energies. You can see that a "small perturbation" in System 1 may not be a "small perturbation" in System 2 because the barrier-energy property that prevents transitions in each system is significantly different. System 2 is less stable than System 1 because it is more likely to transition from state (a) to (b) given equal stimuli.

Figure 2.3 is a classic engineering illustration of a *static equilibrium state* in a simple system. Properties of the two static equilibrium states are determined by the shape of the line on which the balls ride. To carry this analogy further, a similar system in *dynamic equilibrium* may have a line that changes in space over time. For example, if the line is the surface of an irregular ice block being heated. In that case, the jostled ball-on-track system will have properties determining ball position that evolve over time.

Buildings and bridges, if designed properly, can remain in a state of equilibrium for decades or centuries. They are routinely subjected to many types of external perturbations, such as traffic, shifting loads, wind, large temperature swings, and earthquakes. As long as these stimuli remain within the *structural design limits* (analogous to the depth of the gravity well in Figure 2.3), they are considered to be small perturbations that cause small system changes in the form of reversible deformations in building structure. Small reversible deformations ideally do not change system properties. Once the load on a building is removed, any reversible deformations return the system to its original state. Reapplying the load regenerates the same response cycle.

On the other hand, if we increase the loads placed on a building beyond those specified in the design, the perturbations can cause irreversible changes in building-system properties as a result of material yielding, cracking and buckling. In that case, the structure does not return to its exact original state once the load is removed because it was irreversibly changed. Very often small permanent changes accumulate, which lowers the threshold for further change and the system evolves each time a load is applied until it transitions. Small irreversible changes on a system are often described as "wear and tear".

Buildings are subjected to others perturbations from the corrosion of steel beams and columns, cracking of concrete, and rusting of the metal within any reinforced concrete; all system changes resulting from environmental inputs. As long as small perturbations remain small *relative to the design capacity of the system*, we assume they cause small reversible changes in system properties and the whole system remains in equilibrium. Consequently, a few components within a very large system in a highly-variable dynamic equilibrium state can fail (transition) without measurably modifying the functionality of the whole system. This is a *robust system*. In that case, system design capacities at different spatial scales are the criteria for classifying whether a perturbation is "small" or "large".

However, in systems where small changes in structure generate permanent and cumulative changes in system properties, the equilibrium state can evolve instabilities to a point where it becomes vulnerable to sudden transitions following additional small perturbations.

Similar situations take place in the human body every day. Healthy adult human brains operate in a state of rapidly-evolving equilibrium even though a change in system properties is not readily apparent. Our brains constantly undergo internal and external perturbations as we lose and regenerate thousands of neurons every day (internal), and as our senses provide signals to neurons that constantly rewire neuronal connections in response to those signals (external). Each change is relatively small, some changes are reversible, and others are irreversible, especially if the stimulus is periodically reinforced. We say irreversible because information and behaviors that are learned are not easily unlearned. If original ideas are examples of neural-state transitions, we can say with confidence that not all irreversible transitions are to be avoided. Over time, the equilibrium states of a human brain undergo gradual evolution and sudden transitions at multiple scales to ideally progress toward the preferred state we refer to as wise or at least experienced.

People can suffer tumors or strokes that permanently damage neurons in a focal area of the brain (Fig. 2.4). Minor strokes occur when brain cells in small regions are denied oxygenated blood flow for more than a few minutes. Properly treated, many minor-stroke patients fully recover; i.e., they return to a mental state close to the pre-stroke equilibrium because of the brain's enormous adaptive capacity to rewire (neuroplasticity). Another stroke patient, however, might experience permanent loss resulting in dramatic changes in cognitive function, personality, or behavior. These sudden changes in system properties are caused by irreversible perturbations to a brain's neural structure, beyond its compensatory capacity, which result in a transition of the brain system to a new equilibrium state with potentially very different emergent properties such as a change in personality.

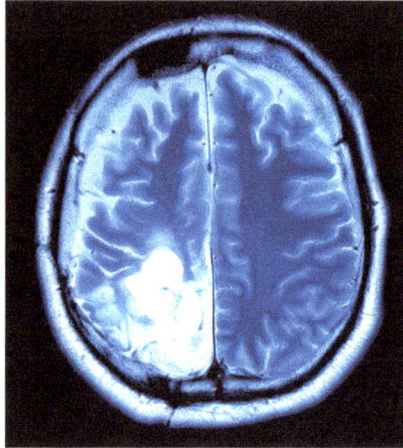

Fig. 2.4   CT scan of a brain lesion. Photo is from Shutterstock.

Much of medical practice can be understood in terms of non-reductionist systems analysis. For example, many healthy-people systems who eat (input) a candy bar will experience (respond) according to the properties of a pleasurable sensation and a temporary energy boost. They may also become slightly heavier; all small reversible responses to the candy-bar stimulus. Hence, the body remains in equilibrium. Conversely, a diabetic person who eats a candy bar without properly preparing can induce a sudden transition into a very unhealthy state. More accurately, transition risk increases over time in diabetic-prone persons who adopt unhealthy lifestyles: poor diet leads to weight gain that, over time, promotes type-2 diabetes where sugar levels in the blood can rise sharply to damage organ vasculature. Accumulating damage continuously modifies organ systems until they become so unstable that eating one more starchy or sugary food triggers a sudden and devastating transition. Cardiovascular diseases, the onset of some cancers, and the consequences of addictive behavior are other disease conditions where genetic predisposition combines with an unhealthy lifestyle to gradually modify the body system, making it more likely that a person experiences a transition to the ultimate undesirable equilibrium state – premature death.

It is important during a slow evolution of a dynamic equilibrium state that an organism remains in equilibrium. Evolutionary changes to a species result from environmental inputs favoring the genetic profile of some individual variants over others. If all environmental perturbations reversibly influence system structures, the system maintains steady-state equilibrium. This is the situation for species that have not changed significantly over millions of years.

Fig. 2.5   A coelacanth fish. Photo is from Shutterstock.

An interesting case came to the attention of scientists in 1938 when they discovered a living coelacanth (pronounced "sèe-la-kanth") fish that was thought to be extinct 65 million years ago at the end of the Cretaceous period in geological history. Fossil remains suggested the coelacanth may be a transitional species between fish and tetrapod, which led scientists to believe it was now extinct. However, in 1938, a live fish matching the distinctive characteristics of the fossil remains was found off the coast of South Africa, and again several times since then at several locations off the coast of Indonesia. It appears that the ancestors of these fish were isolated, living in environments that did not change significantly over millions of years, and so the coelacanth anatomy remained practically unchanged over many millions of generations while fish in other regions evolved anatomically and functionally to adapt to their changing environments. Recent studies and genetic sequencing of coelacanth have shown that it is not related to tetrapods as once thought, and therefore they are not direct ancestors of land animals and human. Scientists now believe that lungfish are more closely related to tetrapods.

The formation of new and distinct species (speciation) is not always slow or gradual. In some situations, significant changes may occur very rapidly. The *Punctuated Equilibrium* theory of evolution [Eldredge and Gould (1972)] describes the process of speciation as long periods of slow and gradual change, punctuated by periods of rapid changes often triggered by significant environmental changes. Evolutionary biology is the study of temporal changes in the hierarchy of organisms that can only be described using non-reductionist methods.

2.4   Types of Systems

Linear systems are a small subclass of natural and manmade systems that have been studied extensively for centuries. A true linear system characteristically generates responses in proportion to the magnitude of the applied stimulus over its full response range. Although no realistic systems are known to meet this criterion, many respond approximately linearly over a small range even if they respond nonlinearly over the full range, as in Figure 2.6. Linear-systems analysis is taught in all colleges of engineering, physical-science, and business programs because this well-defined mathematical theory accurately predicts the responses of a system to "small" input stimuli near static-equilibrium points. Much of the value of linear-systems modeling stems from the intuition they provide us about the internal mechanisms of system functions, more so than from their accuracy at predicting realistic system responses.

All systems respond nonlinearly over some part of their response range, which means they generate responses disproportionate to the stimuli. Extreme examples include chaotic responses and systems undergoing sudden transitions, even though many of these same systems respond linearly to stimuli under other circumstances. Scientists and engineers are taught to carefully evaluate the environmental conditions of a system before adopting the linear classification and relying upon the corresponding analysis.

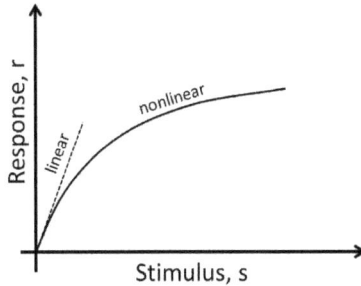

Fig. 2.6   This stimulus-response plot shows how a system with nonlinear responses to stimuli over the full range is approximately linear when the stimulus is small.

Prominent examples of strongly nonlinear-system responses are found in fluid dynamical problems such as those encountered when modeling the atmosphere for weather prediction.

2.5   Sudden Transitions

Although all systems can be found in a state of equilibrium, each system is also capable of undergoing a sudden transition to a new equilibrium state. We described the word "sudden" to mean rapid with respect to the characteristic time scale of changes in a system's natural equilibrium state. In some cases, a transition is very rapid by human standards (measured in years) and in other cases it is slow. For example, light emission is a transition occurring in atoms on a time scale of a nanosecond (10^{-17} yr) as that quantum system seeks a lower energy state. Also some chemical reactions in the body involve transitions of proteins into and out of activation states that occur from seconds to minutes (10^{-6} yr); many economic recessions are system transitions occurring over a few months (10^{-1} yr); global climatic changes, like the onset of a glacial period, can take as little as decades ($> 10^{1}$ yrs) to complete. Each of these transitions is rapid with respect to the age of the equilibrium state of the respective system. Transitions are inherently nonlinear responses of a system to a perturbation, and they often produce irreversible changes in the equilibrium state but not always.

We are familiar from the popular press with the properties of many large complex adaptive systems in dynamic equilibrium; especially economies, brains, and immune systems. If these systems are found in quasi-static equilibrium states, we say they are unhealthy or dysfunctional. We are much happier when we observe the economy as bustling with steady moderate growth; when our brains are continuously learning without undue stress; when our immune systems are being challenged constantly, but moderately, by bacteria and viruses. A potential risk to systems in naturally evolving equilibrium states is their vulnerability to sudden transitions to new and unfavorable states: economies can suddenly fall into recession; brains can succumb to stressful situations leading to sudden bouts of depression or seizure; hyperactive immune responses can lead to autoimmune disorders with a sudden onset of painful symptoms.

We want to avoid these unpleasant transitions even if we are able to recover from them because the systems quite often do not return to their original equilibrium state. Treatment interventions are often sought to transition a system away from an undesirable state to one with more desirable properties, but we are most interested in predicting when a state is nearing a tipping point to either prevent a transition from occurring altogether or to at least guide its trajectory if a transition cannot be avoided. Before we discuss how to detect when a system is nearing a transition, let's consider a few examples.

The simplest system that we have considered up to this point is a piece of rock. The rock is a system because it has interacting components (molecules bound chemically in crystalline formations) that give rise to emergent properties like stiffness. If we know the mechanical properties of a rock in equilibrium, we can predict its elastic deformation (a reversible mechanical perturbation) as it responds to a small squeezing force. If we cyclically apply and release forces with magnitudes near the yield strength of the rock system, the material will eventually develop irreversible property changes – stress cracks. A crack is a small-scale sudden transition in local system properties that affects mechanical properties of the whole. The rock remains in equilibrium even as its equilibrium state evolves with each stress cycle and the whole system is becoming more vulnerable to a sudden transition into rubble. Once it transitions, the shattered rock pieces are smaller and have different shapes than the original. So the post-transition emergent mechanical properties are very different. We can characterize the new equilibrium state from properties of the rubble. What should we look for in the cyclically strained rock if we want to follow the accumulating structural changes in time to intervene and prevent catastrophic failure?

Buildings are larger complex systems than rocks, but our analysis is similar. Building components include support beams, columns, floors, walls, roof, and foundation. Each component connects to another so they interact mechanically, chemically, and thermally over time to give the evolving emergent properties of a building that we see, including its movements in strong wind, the effect of humidity on the materials, and its rate of heat loss in wintertime. The component interactions in a completed building are entirely self-organized according to the laws of physics. However, buildings begin their existence under the strict central control of the design phase before they shift toward self-organized properties as construction nears completion.

Building systems can also undergo sudden transitions. If we apply increasing loads, at some point the materials fail and structures collapse. Or, if some of the columns on the first floor are suddenly removed through explosion, the building will immediately collapse; an approach used in demolition. A collapsing building is undergoing an irreversible transition. The resulting rubble is in a new equilibrium state with properties distinctly different from those of the building before the transition. The rubble is still a complex system because it has components that interact with each other. Small perturbations to the rubble structure cause small changes in that new system, and so the rubble is in equilibrium, just not the same equilibrium as before the collapse. We know this because clearly the emergent properties of the components have changed; i.e., responses to stimuli have changed. What should we be looking at in old buildings so that we can follow structural changes occurring over time? Are there indicators that might be common to all systems?

Complex adaptive systems, like flocks of birds, also undergo transitions. When there are no predators in sight of the birds, their main interaction is to remain in the vicinity of like birds. The main group can break up into smaller groups during foraging and while tending to offspring. Yet, as soon as a predator bird appears, most reassemble suddenly to become a flock, as described in Section 1.2. The appearance of a predator triggers a sudden transition of the bird system with properties distinct from the pre-predator equilibrium state. Once the predator threat disappears, however, the flock transitions back to its original relaxed state of equilibrium. Bird groups are found in at least two distinct equilibrium states – foraging and flocking – where the transitions between states are reversible.

Clearly the properties of each state are very different. Flocking behavior depends on the specific properties of individual bird as well as properties that emerge from the interactions among birds following similar rules. We note that other bird-system transitions in response to attack are possible and are likely to have been implemented at some point in time. Equilibrium-state properties of living systems are forged through evolutionary selection, and so it would seem that the strategy adopted by the flock is the more successful strategy of the many attempted.

Some systems have properties that evolve slowly but in ways that allow for the possibility of sudden transitions. For example, there is evidence to suggest that early solar-system orbits were quite different than they are today. Soon after the formation of the sun, the large gassy outer planets formed. Then as now, most of the water laden materials in the accretion disk lay outside of Jupiter's orbit. The Grand Tack model of the early solar system [Crocket (2015)] posits that Jupiter's orbit varied, allowing it and the outer planets to slowly tack toward and away from the sun as they orbit. This motion hurtled icy asteroids into the inner solar system while robbing Mars of much of its rocky planetary mass. Although the orbits of the gas giants eventually stabilized, early orbital dynamics explain how icy asteroids could have accumulated in the inner solar system during the formation of the inner planets. Without planetary mixing of the rocky inner asteroids with icy outer asteroids, the water on which life depends is unlikely to have accumulated on Earth.

Today the stable orbits of these outer plants help to maintain the equilibrium state of the inner solar system by holding at bay many large, icy rocks from the Kuiper belt just outside Neptune's orbit or further out in the Oort cloud. Similarly, the gravitational field of Jupiter shepherds the remaining rocky asteroids into the space between the orbits of Mars and Jupiter (the asteroid belt), protecting the inner planets from bombardment. So the likelihood of a planetary transition from a large asteroid today is much lower than it was 4 billion years ago in the early solar system. However, there can be gravitational field perturbations in the outer regions that cause icy objects to fall toward the sun; for example, comets. This rare situation has the potential to wreak havoc on any planet in their way. For example, there was a spectacular collision of comet Shoemaker-Levy 9 with Jupiter in 1994. If a large comet collided with Earth, the effects on all life would certainly be transformative. Although sudden transitions in our clock-like solar system today seem unlikely, they are very much possible.

In hierarchically structured complex systems, sudden transitions can occur at small scales that may not induce transitions at larger scales. Consider the ecological disasters discussed in Section 1.10, where explosive rabbit-population growth in Australia decimated sectors of the plant population. If a plant and perhaps a closely-associated ecosystem component became extinct, we can certainly say a local transition has occurred. Yet a robust ecosystem will be able to tolerate lost components if other species can

provide similar functions so that transitions at a small scale do not necessarily propagate to larger scales. Nevertheless, a small-scale component
loss that reduces system diversity can reduce overall stability, making the
system a little more vulnerable to a tipping-point event.

Economic vulnerabilities are similar. One bank may fail (small-scale
transition), which progressively weakens the overall economy in progressive
stages until just one more such event suddenly plunges the economy of a
nation into deep recession. Similarly, a living organism can acquire a localized infection that suddenly and irreversibly transitions cells in a region of
the body. If the immune system can successfully contain the infection to a
small region, the organism can recover; i.e., functional losses are reversible
if cells regenerate or surrounding tissues compensate for the loss. If the
infection spreads, however, then cells in larger regions fail faster than the
system can compensate until a vital organ is lost and the organism transitions into death. And worse, if the affected individual is contagious, an
infectious agent can sweep through and cripple a population. Functional
redundancy is an essential property of any large complex system that is
robust to small-scale component losses. The capacity to buffer component
losses may just delay a transition giving system managers some time to
prevent transitions from occurring or propagating to larger scales. Delays
give economists, engineers and physicians time to monitor their complex
systems and react to looming failure. We return to this point in later chapters. What aspects of hierarchical systems can be monitored if we hope to
detect pernicious progressive processes in time to intervene?

2.6 Uncoupled Representation of System Properties

To begin answering some of the big questions above, we divert our attention for a time to discuss *eigen-analysis*, a central topic of this book.
Eigen-analysis, or characteristic analysis, is an approach used to conceptually break down (analyze) natural and manmade systems into their *uncoupled elemental properties*. Yes, this is a reductionist's tool. An uncoupled
property describes the responses of a system to an input in ways that is
separable from other system properties, and so it is a means for isolating
elements for study without changing the system being studied.

For example, our ear-brain systems naturally perform a type of simple eigen-analysis as we listen to music called frequency analysis. Listening to an orchestra, we note the musicians are generating sounds together that were written to be complementary, and therefore the sounds

are interdependent. And yet, the tones generated are distinctive enough to be separable based on groupings of vibrational frequencies. With a little training and concentration, we can uncouple instrument sounds even though they are played at the same time and written to be merged together. Rather than changing the system, by asking each musician to play at different times and destroying the orchestral effects, we use eigen-analysis to uncouple the sources of the whole system unmodified.

We will see that eigen-analysis is an important tool for defining and modeling emergent properties of complex systems as well as simple linear systems. Our objective in this section is to analyze music chords without the usual mathematical detail.

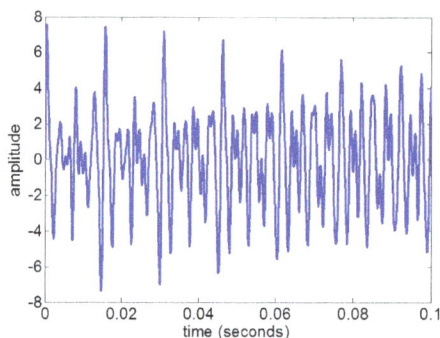

Fig. 2.7   The signal synthesized from a guitar pickup after a C Major chord was struck.

Some musically-gifted people have a natural ability to recognize music chords by just listening to sounds; that is, they do not need a reference-frequency standard for comparison. These people are said to have perfect (or absolute) pitch. Others might develop that skill after long hours of training, but the rest of us require the insights provided by a mathematical analysis of the recorded sounds.

Imagine that we recorded the time-varying voltage signal from a guitar pickup as shown in Figure 2.7. We are told the waveform is from a C Major chord being struck, which we assume is composed of only three notes from the fourth octave played simultaneously: middle C with a fundamental frequency of 261.6 Hz, E at 329.6 Hz, and G at 392.0 Hz. It is certainly not obvious after looking at Figure 2.7 how these frequencies and their overtones mix to give the signal shown, so we break the signal down into component parts to discover the uncoupled properties of the instrument that made the signal.

Reductionist methods teach us that linear systems can be conceptually disassembled into uncoupled properties using eigen-analysis if we understand many details about the system. If we can accurately model instruments like guitars using equations that linearly map applied string forces into output sounds recorded at the pickup, then eigen-analysis can accurately decompose those equations into the uncoupled properties. In practice, we need to make many assumptions that simplify and enable computations. Assumptions are discussed in textbooks and will not be discussed here.

The shapes of guitar-string movements (the standing waves) that are possible are given by the *eigenfunctions* of the guitar system. On a guitar, we often select sine waves at wavelengths determined by the length and thickness of the string as the eigenfunctions. The frequencies of the string vibration and how long the vibrations persist are given by the system's *eigenvalues*. There is one eigenvalue associated with each eigenfunction.

A chord is formed by placing fingers on the fretboard of the guitar as the strings are vibrated. This fret-fingering action temporarily adjusts the length of the strings, which changes the string-vibration modal shapes (eigenfunctions) and the frequencies (eigenvalues) that determine the sound properties of the guitar system at that time. All possible string movements on a guitar at any time are given by weighted sums of eigenfunctions. Since one eigenfunction is uncoupled from the others, the eigenfunction-eigenvalue pairs (sets of these pairs are called an *eigensystem*) are a valuable way to describe the emergent properties of a guitar system. The frequencies that emerge are based on the three notes of the chord (fundamental frequencies) and all the associated harmonic frequencies (overtones).

The eigensystem description is quite general in the sense that it applies to the responses of any stringed instrument. This generality leaves us with many constants in the equations that we must determine experimentally for each situation. This is where fine-scale details of how an instrument was built and exactly how it is played enter the analysis. Such details form the "initial and boundary conditions" of our specific playing situation that enables us to determine the constants. At this point, we can predict what sounds can be produced by specific instruments and how details of the performance and venue influence the sounds we will hear. The predictions made will be accurate if all the assumptions needed for the analysis are reasonably satisfied. Such decompositions of a guitar by eigen-analysis into its uncoupled properties are essential for understanding the nature of the sounds produced. They are not necessary to enjoy music, but they are very useful if you are teaching music theory or building a guitar to have a particular sound.

There are many eigenfunction-eigenvalue pairs possible, in fact an infinite number. The number of *meaningful or accessible* eigensystems (see Section 6.2) depends on the number of *degrees of freedom* in the system: the number of ways a string is allowed to move. Design considerations of the instrument, from body shape to string stiffness, can amplify or attenuate audible vibrational modes in ways that can enrich tonal qualities to make music beautifully complex.

Those who enjoy guitar music are thankful real instruments are not linear systems because of the much wider range of sounds possible with nonlinear responses. (Think of the work of Jimi Hendrix, Jimmy Page and Stevie Ray Vaughan and broaden the system to include electronic amplification and a gifted musician!) To linearize the equations, we might consider restricting the range of operation of the instrument. This action would allow us to compute the eigensystem using standard engineering methods, but those equations would not represent the instrument's true range of properties any more than knowing primary colors can describe a Van Gogh painting. We want to study instruments as they really are, even if they are too complicated to completely model mathematically. The mathematical methods of eigen-analysis that exist today may fail us when analyzing many complex systems. Yet the underlying concept of the existence of an eigensystem representation for any system we encounter remains valid.

2.7   Non-reductionist Eigen-analysis

Mathematically, the eigenvalues of nonlinear systems are not the same complex constants we engineers are taught to expect in school regarding linear systems operating near static equilibrium. Eigen-analysis can be generalized so that eigenvalues become complex functions of space and time. As a complex system, a guitar has uncoupled properties even if we don't now have the mathematical analysis necessary to compute its eigensystem. Just ask a musician listening to music who hears and reacts to the uncoupled properties of instruments in a quartet or a rock band. The trained ear-brain system uses non-reductionist methods to analyze the timing and frequency content of these complex instrument systems in ways we don't fully understand. Current mathematical approaches to uncoupled property determination reach their limits when the system is behaving highly nonlinearly or we do not possess sufficient details about the system. In either case, we might turn to empirical methods where information-based models like those used in machine learning are exposed to a wide range of stimulus-response patterns. This is essentially the method used by experienced musicians.

It is also the approach taken by many craftsmen who make furniture. Consider again a wooden rod as we did at the beginning of the chapter. Its uncoupled mechanical properties can be determined mathematically using eigen-analysis, and it can also be understood intuitively as many experienced furniture-makers do. Wooden rod properties can be decomposed into eigensystems and, because the rod is geometrically simple, we could compute them. Even when the system was too complex to model exactly, the eigensystem exists and may be discovered experimentally.

Say we are interested in the axial deformation properties of a wooden dowel along its long axis. The axial deformation property explains how much force must be applied to shorten the rod a specified amount. We could apply different force patterns to the rod until we found a combination that describes the axial deformation property without changing any of its other properties; e.g., without bending or twisting the rod. This might take some time and special equipment, but if we are successful at applying forces that only compress the rod we are defining its axial deformation property in terms of its experimental eigenstate using *non-reductionist methods.*

The spatiotemporal force patterns help to define the *eigenvectors* (discrete form of eigenfunctions). The magnitude of the energy applied to engage just that one eigenvector defines its eigenvalue. This is a quasi-static mechanical testing experiment where the rod is very slowly loaded while deformation patterns given by very little movement are recorded. In this situation, we can be sure the eigenvalue is a positive real number. A highly compressible rod has a smaller eigenvalue associated with that property than another rod made of less compressible material.

If we soak the wooden rod in water, we will change many of its material properties and hence its eigenvalues. Water softens the rod, making it a little more compressible. Since it requires less energy to compress the rod after soaking than before, the eigenvalue of the axial deformation property is smaller. If we increase the force along the eigenvector associated with the axial deformation property significantly, a dry wooden rod will eventually fail. With high compression, the internal structure of the wood begins to separate and crack, which changes the overall rod axial deformation property. Micro-cracks make the wood more compressible, so the eigenvalue associated with compressibility is smaller. At the instant of mechanical failure, it takes no additional energy to further shorten the rod. If we were monitoring this axial deformation eigenvalue, we would see that its value was decreasing.

When we apply a small compressive force very slowly, the wood responds approximately linearly (in proportion to the applied force), and we can use mathematical methods to compute the constant real-valued eigenvalue and its eigenvector that define that property. As the applied compressive force increases, the response of the wooden rod becomes increasingly nonlinear (disproportionate to the force), the axial deformation property in weaker regions of the wood gradually changes as damage accumulates, and that change can be seen as a now spatially-varying eigenvalue that is approaching zero. This complicated system is not easily amenable to standard mathematical analysis but it can still be investigated experimentally and computationally modeled. As we increase force, the rod suddenly fractures at the instant the eigenvalue for axial deformation reaches zero. At that instant, the two ends of the wooden rod are free to move toward each other with no added energy, and the movement is exactly along the eigenvector of the axial deformation property. Another instant later, splintered wood is lying on the floor in a completely different equilibrium state. Provided we are able to find force patterns that only compress the rod along its long axis and we are willing to load the rod until it shatters, we can discover a tremendous amount of information about this one mechanical property independent of the others *without knowing any details of the materials of which the rod it composed.* Of course, this empirical eigen-analysis does not tell us specifically about the mechanisms of rod failure, so it's of limited design value. It is an indicator of the eigenvalue for a property, which indicated system stability at that instant of time. We expand on this aspect in Chapter 3.

Of course, we might apply different forces that engage different mechanical properties of the wooden rod. We then need to monitor these eigenvalues if we wish to predict failure as those eigenvalues go to zero. The methods described immediately above are straightforward when the system under analysis, a wooden rod in this case, is simple enough to experimentally determine its eigenvectors or eigenfunctions. This procedure is less straightforward in more complex systems.

2.8   More About Eigenvalues

Up to this point, we discussed the eigenvalues of mechanical properties that initially were nonzero. Each describes the energy required to engage one of the uncoupled properties given by the associated eigenvector for a system in equilibrium with its environment. However, systems like a free-standing wooden rod will have at least six initially-zero eigenvalues. These six zero eigenvalues describe the degrees of freedom for *rigid-body motion* of the rod: three translational movements and three rotational. Rolling a rod across a tilted table top requires little additional energy because that rotational mode has a zero eigenvalue, implying that (virtually) zero energy is required to activate the rolling eigenmode.

Because rigid-body modes have zero eigenvalues, they have no effect on the equilibrium state of the rod. In the following discussions, we may ignore them unless they are required for a full system description. Once the rod is glued into a chair and becomes part of the chair system, we have eliminated the rigid-body modes. With six fewer degrees of freedom, the rod has six fewer eigenvalues and eigenvectors, which is a good thing for anyone hoping to safely sit in the chair!

This discussion could be a little esoteric at this point. We ask the reader to be patient until we return to the idea of rigid-body modes in later chapters, where the concept will be shown useful for predicting system behavior during transitions.

In summary, eigenvectors represent uncoupled properties of a system. The corresponding eigenvalues tell us quite a bit about the importance of those properties in defining system behavior. When we observe if an eigenvalue is a real number or a complex function of space and time, we obtain information about the ability of environmental forces to activate a property. This is true of wooden rods that are simple enough for us to calculate the eigenstates using traditional mathematical methods. Yet it is also generally true for all systems, whether or not we know all the properties or can compute the system eigenstates. The value of these ideas is for understanding those complex systems too complicated to model mathematically. Eigensystems can sometimes be found experimentally. Based on the above discussion and prior work [Ghaboussi (2012)], we postulate the following.

2.9 Unifying Principles Governing Properties

Proposition 1. *All complex systems have emergent properties, and uncoupled forms of those properties are represented by the eigenvalues and eigenvectors of the system.*

Proposition 2. *All systems are capable of remaining in equilibrium states where small perturbations result in small changes in the system.*

Proposition 3. *All systems are capable of undergoing sudden transitions from one equilibrium state to another. When the real part of at least one of the non-zero eigenvalues approaches zero, small perturbations cause large changes in system properties indicating the system is about to transition to another equilibrium state.*

Proposition 4. *The initial path of the system's transition to a new equilibrium state is along the linear combination of eigenvectors for which the real parts of the corresponding eigenvalues approach zero.*

The authors have had many discussions about which events constitute a transition, and the matter is not entirely resolved in the sense that we agree in each situation. Nevertheless, we do agree that for a complex system to transition, a small perturbation generates a large response, an eigenvalue of the system has approached zero, and the system exists in a different equilibrium state after the transition than before, which means it has different properties.

These four propositions offer a new way to think about systems in general, especially about the all-important sudden transitions that cause a system to undergo major changes. Eigen-analysis enables a non-reductionist assessment in the sense that we do not need to understand every detail of a system, only its uncoupled properties, which we can measure experimentally in some cases.

The exciting feature of these propositions for us is that they offer insights into systems now deemed too complex to analyze. For example, based on these propositions we hypothesize that the real part of at least one of the non-zero eigenvalues of North African and Middle Eastern societies was approaching zero at the time of the Arab Spring transition in 2011. Even though we currently have no idea how to compute the eigenvalues of a society, we know they are systems with emergent properties and hence they have eigenvectors and eigenvalues. These eigenvalues will change as the societal-system structure and its properties evolve in time. The gradual changes within societies resulting from improved healthcare produced a younger population that is better educated and has wide-spread access to information technology via the internet. These internal changes caused the real part of some of the eigenvalues associated with political stability to approach zero; their equilibrium state had evolved to a point where a small perturbation was able to cause major changes – a sudden transition. Self-immolation of Mohamed Bouazizi, a street vender in Tunisia, was that perturbation. It led to mass protests that toppled the governments in Tunisia, Libya and Egypt and civil wars in Syria and Yemen.

Similarly, we can state that in September of 2008 the real part of the some of the eigenvalues of the financial market were approaching zero and the US economy underwent a sudden transition into recession. The small perturbation that triggered the transition was the bankruptcy of the Lehman Brothers investment bank. We believe non-reductionist eigen-analysis can help us to understand what changes occur within very large complex systems that lead to their instability. This approach points to methods for predicting system transitions provided we can link system features to eigenvalue changes.

In Chapter 4 we graphically describe simple systems for which we have computational models and can find the eigenstates. We will demonstrate that when these systems are about to undergo sudden transitions to new equilibrium states, at least one of their eigenvalues approach zero. In Chapter 6, we will describe generally stable systems that can have many eigenstates with eigenvalues at or near zero; these are systems in a state of *self-organized criticality*.

The behavior of eigenvalues during a transition as discussed above is a direct consequence of every system always having properties and existing in an equilibrium state. Although the dynamic equilibrium state we describe is not the static state defined in classic thermodynamics, there is a sense that a system property is stable because of counterbalancing internal factors at work among the component connections in a system; e.g., think of the tensegrity model described in Section 1.7. Hence, if some components of systems with offsetting opposing factors are minimized, i.e., one or several eigenvalues approach zero, then there is virtually no energy required to engage the associated eigenvectors. If you are sitting on a three-legged stool and one leg suddenly breaks, you fall according to the motions dictated by the eigenvectors associated with the eigenvalues that suddenly went to zero. Expanding this logic to larger more complex systems is a logical and, we think, reasonable assumption if the four propositions above are unifying principles of all systems.

Chapter 3

Knowledge Discovery

The description of system properties in Chapter 2 ends with a list of four propositions describing unifying principles that we believe apply to all systems. Unifying principles are essential for building an analytical framework, but they do not directly point to procedures for characterizing and predicting system behavior. In considering various analyses that might be used to study systems, we found it important to examine the nature of knowledge, how it is discovered, and how these methods might be used to further the exploration of complex system behavior.

3.1 Scientific and Engineering Questions

Science is a process that describes systematic efforts by humans to accumulate, analyze, synthesize, and disseminate information leading to an understanding of the structure and behavior of the phenomena we perceive. Although the term frequently refers to the physical sciences where the topics are matter and energy in its many forms, the value of science to all human endeavors is to offer general analytical approaches, via the *scientific method*, that knows few topical bounds. Knowledge is discovered through insights obtained from *information* that serve to augment and correct the existing knowledge base. Unlike information, knowledge conveys a sense of value through utility; and yet knowledge remains in draft mode, waiting to be edited after thoughtful consideration of new information that generates another perspective among the facts that imbue value. Knowledge is held individually and collectively as an ever-changing collection of information, principles, and skills that thrives only if it is exchanged among the populace.

Science provides a common language and approach to weaving sensory experiences and abstract idea threads into a knowledge network. We tell students eager to make contributions to the network that if their ideas are to become knowledge, those ideas must be clearly communicated and analyzed in the context of what is currently known. Thus, we record new information and analysis along with existing models and theories so that others may consider the ensemble information and build the concepts further. Circulation of information and debate on its value are essential ingredients for knowledge vitality. Knowledge demands frequent affirmation through experimentation followed by logical consistency checks. Civilization thrives as knowledge circulates rapidly among people just as the economy thrives when money circulates.

Knowledge is a complex adaptive system into which information gleaned from human perception and imagination is input. Science is an analysis/synthesis/filtering process for converting information into knowledge. Knowledge resides in its system components – human brains augmented by literature and other forms of information storage – as it circulates to remain relevant. The system output is our individual and collective world views; i.e., the structural and functional models of how we believe everything works, which guides society as a whole and each of us on our journeys through life.

Because knowledge is born from human observation and analysis, the complicated interface between actual world events and human perception via senses/cognitive functions plays a prominent role in determining variability in the dynamic equilibrium state of knowledge systems. The variability in perception and information processing among individual humans generates variability within the knowledge system. That variability is healthy when expressed as diverse opinions and interpretations of information and ideas. The dynamic nature of the equilibrium state of knowledge may frustrate people unaccustomed to valuing diverse world views equally; it seems that every other week we hear that some practice thought to be harmful (healthy) is actually healthy (harmful) for people or society. Oscillating conclusions must be expected by the public when simple summaries of a complex and incomplete knowledge base are requested and offered. Some people assume that knowledge, once vetted, is objectively held in a static equilibrium state. Not so! Relevant knowledge cannot be static. Used in a broad sense, *culture* is the name given to the dynamic knowledge system that surrounds us, the ethos that guides our actions. Our local culture is highly dependent on the range of environmental inputs to which we are subjected. *Education* is the training we receive on how to evaluate information to which we are exposed, beyond our instinctual and tribal tendencies, and to avoid becoming insular and extreme in our world view.

We only have educated *reasoning* as the guardian to keep riffraff from entering the hierarchies of our knowledge depositories. The guardian decides whether to accept new knowledge based on current knowledge; this is a naturally self-regulated process with little or no central control. If great pains are taken to exert control, that control is often temporary and ultimately sterile with respect to generating new knowledge. The guardian is not trustworthy without proper training, and even so it makes the mistakes of any self-regulated process. The value of a well-connected and educated population, where each person is trained to independently scrutinize for consistency every information morsel that presents itself at the gate, is monumentally critical to the quality and utility of humanity's collective knowledge. This base, after all, offers the only tools we have for making the tough decisions that society demands we make. One of our goals with this book is to "stir the pot" of knowledge within a distribution of interested readers. The hope is that greater awareness of the issues and challenges brought by living in a world of complex systems will breed innovative thinking and, if we are fortunate, new solutions to ongoing problems involving their behavior.

In the broadest sense, *engineering* is the application of knowledge to solving specific problems involving systems. The differences between science and engineering are related to the different questions each asks. Science asks how or why something works the way it does; it is the search for mechanisms underlying an action. Conversely, engineering depends on that knowledge to address problems related to the design, construction, evaluation, and maintenance of the systems involved with an action.

For example, a scientist might discover the principles of friction to explain how one object slides across another, while an automotive engineer might use those details and many others to design a car that maintains driver control under icy conditions. Another scientist might work to discover the influences within organisms can cause a cell to transition into a cancerous state. An engineer will apply knowledge of cell mechanisms to design diagnostic and therapeutic procedures for detecting and treating cancer in patients. A scientist often looks closely at system details in search of the principles behind an action, while an engineer often must look at all system scales to effectively apply the principles as s/he solves specific problems.

Each professional person in academia, business, industry, medicine, and other organizations, depending on the situation, must operate as a scientist and as an engineer to be effective. Obviously, we are using these terms very generally.

"Scientist" and "Engineer" are labels that can apply to many paired fields of study, even those outside technical areas. One term or another will apply depending on the questions being addressed. The "scientific fields" labeled chemistry, biology, neuroscience, psychology, fine arts, mathematics, and physics are matched, respectively, by "engineering fields" labeled chemical engineering, medicine, psychiatry, social work, architecture, computer science and civil, electrical and mechanical engineering. At many universities, engineering departments have the words "science and engineering" in their names to reflect the value they place on multifaceted approaches to research and that those influences are fully integrated into their educational programs. To be effective professionally, scientist-engineers learn very quickly to stay well connected to the vast knowledge network.

One may hear opinions that science or engineering is more important or more valuable than the other; such discussions are nonsensical hubris since each depends on successes of the other. For example, the development of advanced microelectronic and nanotechnologies (engineering) empowered the revolution now known as cellular and molecular biology (science), which in turn motivated new technologies in an upward spiral of activity (see Fig. 3.1). Similarly, computer engineering and information sciences explore different worlds and yet are entirely codependent. The late-career years of 19th-century chemist-microbiologist Louis Pasteur is a great example of one person who dug deeper into fundamental microbiology research in order to advance his overarching applied objective of making drugs that reduced suffering from pathogen exposures. He changed questions rapidly, sometime asking as a scientist and other times as an engineer, to achieve his real goal, which was solving big societal problems.

We posit that the study of complex systems as an entity similarly requires the tools of the scientific and engineering worlds to be coupled, but perhaps in ways that have yet to be imagined. We want to understand the mechanisms that generate output responses from the inputs to complex systems (scientific questions), and we need to accurately predict the evolutionary paths of equilibrium states, especially when they increase the likelihood of sudden transitions (engineering questions).

Trained scientists studying a complex adaptive system often adopt reductionist approaches that ask "What are the basic components and

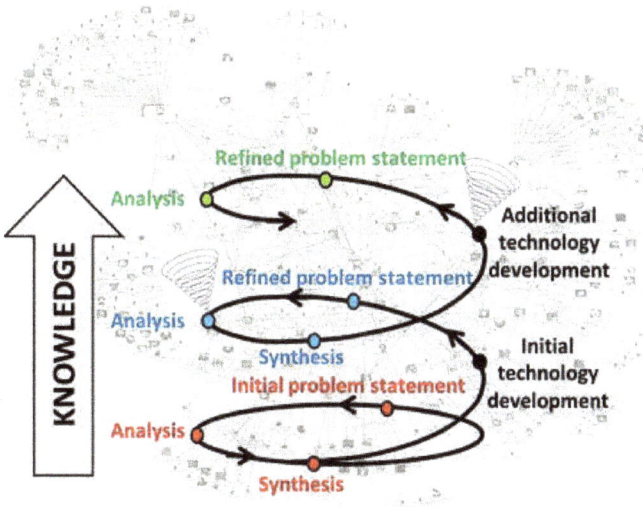

Fig. 3.1 The illustration shows a spiral pattern of research and development that spawns new knowledge within the knowledge network. Scientific and engineering aspects arise depending on the questions asked. Technology development falls within the implementation phase of the cycle. The small conical helices shown indicate that new research is spun off, which adds benefit to investments made in highly innovative ideas. Investments are maximized when new knowledge is clearly and widely disseminated.

interactions of systems that generate the emergent behavior observed?" Yet, once we realize that emergent properties are encoded in component interactions at potentially all scales of the networked components, we then need to wonder how close we should look at the system to find fundamental properties? If we knew all of the details at one scale, we still might not know what happens at another. Something is created as system components assemble. For example, knowing the detailed behavior of individual muscle cells is not sufficient to predict how much weight you can lift. Knowing how viruses infect cells to multiply their numbers is not sufficient to predict how quickly the virus will spread in the population. How should we connect investigations made at various scales of a system? Even if we could understand the details of an adaptive system at one instant of time, its response to exactly the same input at a later time can be quite different. So we need more than reductionist tools for this analysis. We must learn to synthesize the methods of reductionist and non-reductionist analysis to understand and predict system behavior. The history of science offers some clues on how to proceed.

3.2   Early Forms of Knowledge Discovery

Early in human development, at least tens of thousands of years ago, humans were all engineers trying to solve a long list of specific survival problems. We discovered that hunting and gathering as a group provided participants advantages over those that hunted alone, which allowed the group to flourish even in harsh or threatening environments. Survival was at the center of every activity. Groups assigned tasks based on individual skills as they acquired food, avoided being eaten by predators, raised the next generation, and dealt with harsh geographic and climatic conditions. Knowledge discovery occurred spontaneously as humans, like all animals, used instinctive decision-making to search for solutions to survival problems. Without knowledge repositories, e.g., books, human knowledge was passed to others through apprenticeships and storytelling.

Knowledge discovered from trial and error is a form of stimulus-response learning used by all animals. If you survived an event to remember an experience as positive, then you used the solution again and again to eventually develop a skill. This could be taught to others, and in this way a collective knowledge base is built to be passed to the next generation. The group valued observant members who could recall past successes and connect important aspects of that information, which formed the knowledge network shared by all in the group. Knowledge has been a complex adaptive system from the beginning of civilization. Humans who learned how to make and control fire, make weapons and other tools, and organize the group effectively were more likely to survive and prosper. Leaders of successful groups motivated searches for solutions to survival problems that, if reinforced, resulted in general knowledge to be used to solve many problems. If fire could be controlled for cooking meat, it might also be useful as a defense weapon against predators at night, and, if that worked, maybe it would be useful as an offensive weapon that eliminated future threats. So the earliest approaches to problem solving were highly pragmatic: solve the problem at hand to develop a skill that enables the same problem to be solved more effectively in the future. This is knowledge development through survival engineering.

The most clever problem solvers generalized their solutions by keeping an eye out for different problems where the same or similar solutions also fit. If that worked, then a general principle was formulated that seeded solutions to still more problems. Without formal methods for learning and a reliable recording of successes and failures, knowledge generation must have been painfully slow. Knowledge is like money, the more you possess the more and faster you can make it accumulate.

We see knowledge discovery through the process of trial and error repeated throughout human history. The magnificent Egyptian pyramids in Giza have remained stable for thousands of years. Setting aside the challenges of the building process, how did designers know how steep to build the sides of pyramids so they remained stable? If a slope is too steep, the sloped wall fails by slumping and spreading. Slumping is when sheets of material on a sloping surface break away from the bulk and slide downward. Modern civil engineers are very familiar with slumping and routinely design structures like levees to avoid slump failures. A gentle slope, one much closer to horizontal than vertical, is always stable but not always effective or economical. As a slope increases toward vertical, it becomes increasingly unstable. At what angle does a slope change from stable to unstable? There is historical evidence that knowledge of how to build large stable pyramids evolved gradually through trial and error until a principle was formed and recorded. Trial-and-error knowledge discovery was also useful in the construction of cathedrals, bridges, buildings and more generally in the developments of large-scale agriculture, government, and warfare.

Success at surviving led to at least two societal benefits. The first benefit was greater confidence and skill at abstracting principles from successful experiences that could be applied to other problems. The second benefit was relief from constant attention to survival activities that generated leisure time to further develop new intellectual skills. People had time to follow interests wherever curiosity led and the skill for pursuing knowledge without a specific purpose – just for the sake of acquiring knowledge – and for all the satisfaction, power, and fortune that the new skills brought. Free time led to developments in mathematics, agriculture, philosophy, astronomy and government. Development of these skills and the methods of scientific inquiry started mainly along the Nile, Mesopotamia and Ganges Rivers, where efficient large-scale agricultural sciences provided a high quality of life and more free time for scientific inquiry to satisfy individual curiosity.

3.3 Discovering Knowledge About Systems Today

Racing ahead to today, mankind's motivation and methods for pursuing knowledge have evolved. A strong motivating factor for learning is still curiosity, an attribute encoded in our genes but one that requires nurturing to be fully exploited. Curiosity is reinforced in a person when they realize that new knowledge can be generated from acting on information. The life-long need to learn experienced by many people may have genetic roots because of its survival advantages, but learning is itself learned behavior that can border on compulsion for the most erudite. In modern times, with the development of a global society, knowledge discovery has shifted largely from a survival-driven to a curiosity-driven motivation, in which scientific discovery has blossomed to the great benefit of society and science's pragmatic partner, engineering.

Curiosity drives scholars in pursuit of knowledge to "dig deep" for answers to questions about why systems respond the way they do. That is, to understand systems, we must first know what is fundamental about that system. Physicists are the most reductionist of scientists, and with good reason given their long history of successes in searching for system fundamentals. In searching for the fundamental laws of nature on which other fields of study are built, physicists found hierarchies within matter structures built from elementary particles that interact according to four fundamental forces.

The reductionist assumption is that nature at its core is composed of elementary particles that move and cluster according to fundamental forces, and that all the information required to understand the behavior of large systems built from these parts are somehow encoded in these elementary components. Richard Feynman stated in his popular *Feynman Lectures on Physics* [Feynman *et al.* (1964)] that "there is nothing that living systems can do that cannot be understood from the point of view that they are made of atoms acting according to the laws of physics." This reductionist dream has suggested a Theory of Everything (ToE) might be possible. It would form the ultimate basis for future scientific and engineering research; the essence of any system would always boil down to applications of ToE principles to understand all system properties. How great would that be!

Fifty years ago, Victor Weisskopf suggested that physics offers two approaches to knowledge discovery, *intensive research* is the reductionist search for the fundamental laws of nature, and *extensive research* is the non-reductionist effort to explain phenomenon through applications of the fundamental laws [Weisskopf (1965)]. Essentially, intensive research focuses on analysis (decomposition of systems to form knowledge of elementary components and interactions) with the goal of understanding the fundamentals of nature. Extensive research focuses on synthesis (conceptual reconstructions of specific systems from knowledge of the elementary components of nature) with the goal of predicting and controlling system responses. This pair of approaches helps scientists and engineers formulate the questions they research. If strict reductionism can show us how nature operates, that all knowledge of systems can be derived from a synthesis of elementary laws, then scientists should keep digging deeper for fundamental knowledge and engineers need to learn to effectively synthesize that knowledge into comprehensive descriptions of systems. If this is true, then we need to all keep doing exactly what we do now.

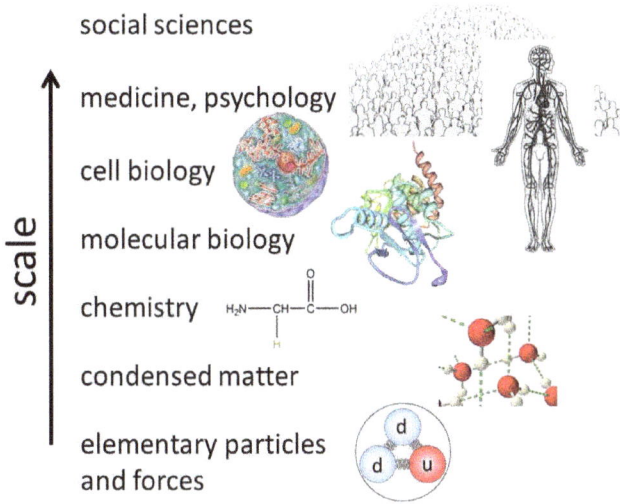

Fig. 3.2   Hierarchy of structure in the human system.

A few years later, Philip Anderson published a remarkable paper entitled "More is Different" [Anderson (1972)]. This paper is easy to find on websites, and so we urge all readers to read it despite its technical elements. Anderson acknowledged the importance of reductionism in knowledge discovery, but he also challenged the viewpoint that emergent properties of a system could be synthesized through a reconstruction of elementary components. Anderson stated, "The constructionist hypothesis breaks down when confronted with the twin difficulties of scale and complexity." He noted that as components are combined to form systems at increasingly larger scales, new fundamental properties emerge that are not found at the more elementary levels. The only conclusion that makes sense is to reason that fundamental knowledge must be gleaned from intensive research at all scales.

Anderson used a biological example similar to the one diagrammed in Figure 3.2. Atoms are assembled from particles that assemble to form molecules that form cells, etc., up to populations. From the formation of systems at each level, properties emerge that did not exist nor can they be explained from properties at the lower levels. Just as elementary particles and forces determine the fundamental properties of the components of matter, so too the emergent properties at each level of structure are fundamental to the behavior of a system at that level. In essence, something fundamental results from the component interactions within a system.

Anderson explained his position by stating it is "essential to realize that matter will undergo mathematically sharp, singular phase transitions to (equilibrium) states in which the microscopic symmetries are in a sense violated." He was describing the emergence of properties with increased complexity.

Symmetries refer to the different perspectives from which system components appear to be the same, e.g., molecules in a gas. In many ways, physics is the study of symmetries at different scales. As a simple example, the Earth is rotationally symmetric because it looks the same every 24 hours to someone sitting on the Sun. Also, hands are reflectively symmetric because your right palm viewed in a mirror looks the same as your left palm viewed directly (try it!).

To gain some appreciation for the importance of symmetries, consider the simple molecular example diagrammed in Figure 3.3. Individual oxygen and hydrogen atoms, which are each highly symmetric, self-assemble for energy reasons to form less-symmetric, tightly-bound water molecules (Fig. 3.3a). It is really the electronic and rotational-vibrational energies that define molecular symmetries more than their shape, but shape works to understand the basic concept. Now assume a container of water vapor is being cooled. Weak (hydrogen) bonds begin to form between them as the molecules begin to slow their frenetic thermal dance until water suddenly changes phase from a gas to a liquid (Fig. 3.3b). Order within the water molecule system increases at the expense of molecular symmetry. Simply, as molecules attach to each other, there becomes fewer ways to view these components as the same, either structurally or energetically. Continued cooling further increases molecular order until the liquid crystalizes into a solid and the hydrogen-bond density rises to four bonds per molecule (Fig. 3.3c).

This temperature reduction we have been discussing can be applied very gradually, but the change in water-molecule structure is not gradual.

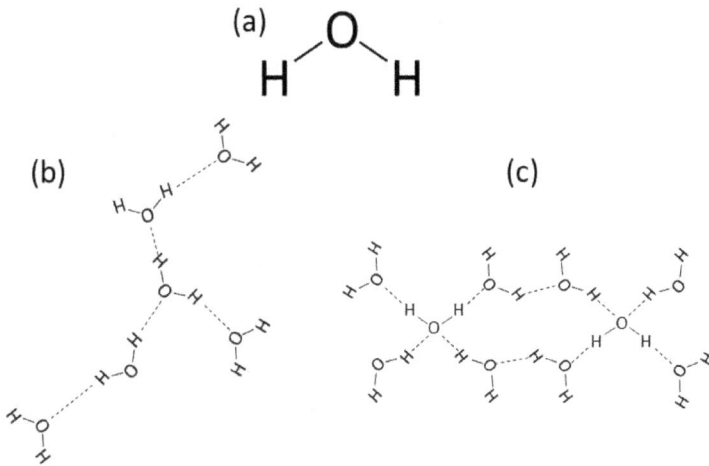

Fig. 3.3 (a) A water molecule has the shape of a bent dumbbell; the two hydrogen atoms H are tightly (covalently) bonded (solid lines) at an angle to an oxygen atom O. (b) In liquid form, water molecules are weakly bonded (dashed lines are hydrogen bonds) to neighbors. (c) In solid form, four water molecules are bound with hydrogen bonds to each molecule to form the regular patterns of ice crystals.

At certain values of the falling temperature, there are "sharp, singular phase transitions to states in which the microscopic (point) symmetries" are spontaneously broken. Symmetry breaking is a passive system transition where individual components change from being highly symmetric and disorderly to being less symmetric with self-assembled group order. Understanding molecular symmetry has provided scientists with a wealth of information, such as being able to predict phase-transition temperatures (e.g., the freezing point of water) and the molecular structures responsible for these emergent properties. Like Anderson, we believe these ideas about physical processes apply broadly to all systems.

The phase transitions in biological-system assemblies that Anderson described (see Fig. 3.2) are much more complicated processes. Here, disordered organic molecules actively self-assemble under DNA guidance with the assistance of energy from the surrounding environment to form macromolecules and more. The reduction of molecular symmetry at each structural level is matched by an increase in information as new structures give rise to emergent properties. Enzymes provide an example of the emergent properties of assembled biomolecules. These are small 3-D molecules manufactured in cells that have the amazing property of dramatically speeding up chemical reactions to make essential processes like signal transduction and regulatory functions in cells possible. Their catalytic functions are emergent properties that are not explained by their components. Anderson's insights about emergence in complex systems are being rediscovered in new contexts these many years later.

As system components assemble, the loss of symmetry is matched by an increase in information. Anderson argued that different fundamental properties can appear at each scale in the assembled hierarchy. While there is no doubt that science has developed methods for reducing systems to their elementary parts so we can understand their functions, we worry about the limitations of synthesizing complete systems in models from these disassembled parts. Essential properties could be missing in such models. We believe there is a need to develop new methods for discovering the emergent properties of complex systems at different scales.

3.4   Eigen-analysis for Discovering Emergent Properties

In light of Philip Anderson's insights about the emergence of fundamental properties, let's return briefly to the discussion of the wooden rod in Chapter 2. We noted that a wooden rod has nonzero eigenvalues that characterize the quasi-static mechanical properties important to the wooden rod's performance as a chair-system component. These initially-nonzero eigenvalues indicate the energy required to deform the rod in the pattern of their corresponding eigenfunctions. Before the chair is assembled, each rod also has six initially zero-valued eigenvalues describing the rod's freedom to translate and rotate freely in space. These movements require essentially no energy; in fact, a cylindrical rod will roll across a table with the slightest flick of a finger. The initially zero-valued eigenvalues are not thought of much because they have no bearing on the performance of the rod as a component of a chair; they disappear[1] once the rod is glued into place. Yet, if a heavy object is placed on the chair that is beyond the capabilities of one or more of the rods to hold this weight, the chair will fracture. From a system's perspective, one or more of the chair-system eigenvalues will approach zero so that the chair system will transitions from its furniture equilibrium

---

[1] Actually, they become part of the chair's 6 zero eigenvalues.

state into a wood-scrap equilibrium state. All this was explained in Chapter 2.

Comparing the assembly of wooden rods into a chair with the assembly of water molecules into a block of ice and organic molecules into an organism, we see that phase transitions remove the initially zero-valued eigenstates from the system components. A bond formed between water molecules is equivalent to a wooden rod being glued into a chair. Post-transition, both systems develop eigenstates that characterize emergent properties of the new assembly. The new eigenstates define the newly emergent properties. This observation is likely to be true of all complex systems.

Through these examples we are suggesting that eigen-analysis be used as a framework for discovering fundamental knowledge about the stability, evolution, and transition of emergent properties in all systems. We discussed the loss of zero-valued eigenvalues during assembly-type phase transitions; e.g., the processes of wooden furniture making, liquid water freezing, or gene expression leading to protein formation. We also described situations where systems are changed by their environment in ways that cause nonzero eigenvalues to approach zero, which then suddenly initiates a disassembly-type phase transition; e.g., the processes of slowly increasing the compressive stress on a chair or block of ice until there is a sudden fracture and a transition to a less ordered state. There is another phenomenon involving eigenvalues that is valuable for engineers seeking techniques to analyze and control large complex systems that we now describe.

Bridge Example The failure of the Tacoma Narrows suspension bridge in 1940 (Fig. 3.4) is just such a case. Its lessons have been taught to generations of beginning engineering and physics students because of the principles it demonstrates. First, a little background. Short-span bridges are fairly stiff and less susceptible to wind forces. Suspension and cable-stayed bridges that are used for longer spans are more flexible and more susceptible to wind forces. Principles drawn from eigen-analysis are used by engineers in designing reliable suspension bridges.

Suspension bridges are flexible and so they can move with environmental forces including traffic, wind, and earthquakes. Like guitar strings, they have natural modes of vibration (dynamic eigenfunctions) that are determined by their shape, materials, and support structures. Currently, bridge designers often determine the eigenstates of the bridge plan so they can predict if a design will be able to resists undesirable movements along any natural mode that might be stimulated by environmental forces. There

Fig. 3.4   The Tacoma Narrows suspension bridge fails in high winds on November 7, 1940. If you look closely you can see evidence of excitation of twisting eigenfunctions in the bridge deck as the bridge collapses. With permission from University of Washington Libraries, Special Collections, UW21422.

are notable examples of bridge failures in the 19th and early 20th century where it was thought that movements of soldiers marching in step across the bridge could match one of the bridge's natural vibrational frequencies. Resonance is the condition where small periodic forces, such as those applied by an army marching in time, generate increasingly larger-amplitude movements if the applied force frequency matches a vibrational frequency. Once large bridge movements cause the failure of some of its components, the whole bridge may fail in response to the gravity field. It is for that reason that troops walk rather than march over bridges. However, the Tacoma Narrows Bridge disaster was caused by a different and much more interesting phenomenon.

This bridge was a victim of high winds velocities for reasons illustrated in Figure 3.5. Wind traveled down the narrows to blow at the side of the

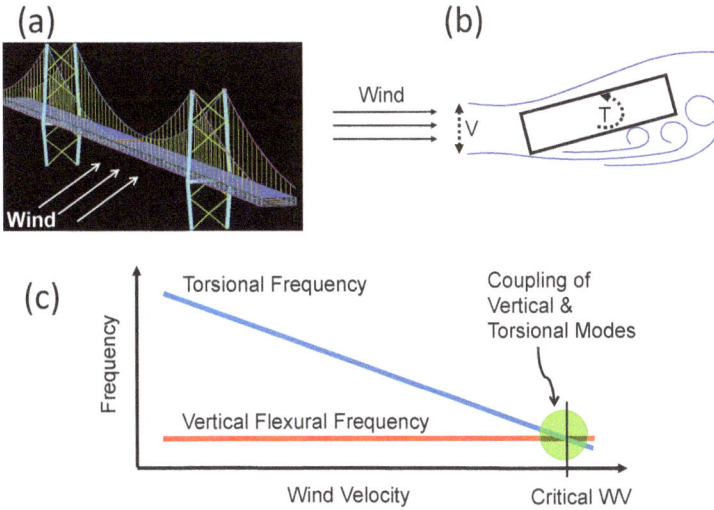

Fig. 3.5 The coupling of two eigenfunctions at the critical wind velocity leads to the collapse of the Tacoma Narrows.

bridge (Fig. 3.5a). The air flow across the deck (Fig. 3.5b) caused a twisting or torsional motion (T) called aero elastic flutter. Torsional motion is a manifestation of one of the many eigenfunctions of the bridge system. A second eigenfunction responsible for vertical flexure (V) is typically not influenced by wind. Vertical flexure causes the deck to move up and down. Since T and V modes are eigenfunctions of the system they describe uncoupled properties of the bridge, and since they have different eigenvalues the modes are independent. Designers rely on modal independence to maintain structural stability.

The two eigenfunctions each have an eigenvalue that depends on the vibrational frequency of that mode. When the wind velocity is low, the torsional and flexural frequencies are different (see the graph of Figure 3.5c). Consequently, the eigenvalues are different and the two vibrational modes are independent. As wind velocity increases, however, the frequency of the torsional mode decreases due to flutter such that its eigenvalue approaches that of the vertical mode. At a critical wind velocity where the two frequencies are equal, their eigenvalues equate and the two modes are linked. The linkage means that large torsional-mode motions can now be transferred to vertical-flexure mode motion, and suddenly the bridge deck is not only twisting but galloping up and down. That occurred on the Tacoma Narrows Bridge in 1940 until the excessive motion caused the bridge failure. We know the eigenvalues for the two modes did not go zero, even as the bridge collapsed, because the T and V modes can be seen continuing in Figure 3.4 as the deck is falling into the water. Designers now know they can change the shape of the deck and rearrange support structures to effectively extend the critical wind velocity to much higher values to avoid the modal linkage. This example teaches us that linked eigenfunctions exist with potentially large consequences for system behavior. Two modes link when their eigenvalues become equal because they form a subspace where together they behave as an independent eigenfunction of the system.

Let's compare the eigen-analysis of a guitar string from Chapter 2 with the bridge example above. Building a guitar, designers select component materials and shapes and assemble them into structures that determine all the sound properties possible – the eigenfunctions (possible standing-wave vibrational modes) and eigenvalues (modal frequencies) of the system. Playing the guitar, musicians place fingers on the fretboard to temporarily modify system properties, so when they strum the strings (apply input to the system), they excite some of the possible modes (eigenfunctions) that produce the sound we hear (output of the system). Sliding fingers down the fretboard toward the guitar body modifies tones by shortening the fundamental-frequency wavelength of string vibration and the associated harmonic frequencies. Vibrational modes couple to material and geometric properties of the instrument, the musician, and the venue to determine the eigenvalues of this system. The engineering goal of guitar playing is to form fingering actions that activate specific eigenstates with the appropriate timing. The musician's input reversibly modifies the system itself: fingers on the fretboard modify the eigenstates and the string amplitudes induced by the pick determine how nonlinear the system response will be. If a guitar string breaks while being played and just one eigenvalue goes to zero, at the instant of the break, the string assumes the shape of the associated eigenfunction before collapsing, like in the bridge example of Figure 3.4. If two or more eigenvalues of the guitar become equal at some time, then activating one of these vibration modes can transfer energy to the others whether or not that action is intentional.

Building a bridge, designers also select component materials and shapes and assemble them into structures that determine all possible bridge movements. Uncoupled properties of the bridge are described by the eigenstates of the system. Normally, environmental inputs (wind, traffic, etc.) load the bridge and induce small movements according to eigenfunctions within the safe-operating design limits. This system is intended to resist distracting bridge motions for drivers that might be caused by the environmental inputs. Aging or damaged materials can lead to an evolution of the equilibrium state of the bridge. If these changes allow one or more of its nonzero eigenvalues to approach zero, then the slightest environmental energy will cause undesirable movement that could result in a bridge collapse. However, if the evolving equilibrium state and/or shape of the bridge allow two or more eigenvalues to become equal, vibration modes are no longer independent, and, in the case of a suspension bridge, linked modes can be dangerous.

The eigenstates of a guitar string are fairly easy to compute by first solving a well-known partial differential equation. The eigenstates of a bridge are more difficult to compute but it is still possible with simplifying assumptions. Does the guitar-bridge analogy hold for more complicated systems? We think it does under eigen-analysis, which follows principles governing all systems.

Large complex systems, where connections among the components are known, can be represented by a graph like that seen in the background of Figure 3.1. Graphs are drawings of a system that specify the relationships between its components. These relationships can be summarized with a mathematical structure called a matrix so that computational methods can be applied to approximate its eigenstates. This general approach was used with much success by Schilling and Palsson to identify active pathways within the many large and coupled networks of biological cells [Schilling *et al.* (2000)]. Eigenfunctions define possible pathways through convoluted cellular networks that enable the broad range of cell functions needed to sustain life. Some eigenvalues are related to activation energies. If an eigenvalue goes to zero, a cell may not be responsive to a mode through the network or it may only respond to that mode. This information is valuable to know when discovering how cells function or when engineering drugs that inhibit or promote those functions. Can cellular eigenstates couple like those in mechanical systems? We don't know for sure but the concept is rich with possibilities. This topic is explored in Chapter 5.

This is an engineering description of cellular functions known as *systems biology*. It is an example of how engineering thinking has permeated other disciplines. The hope is that systems biology can form a basis for promising future medical technologies, which is possible only if each cell network can be accurately mapped. Cell networks vary among individuals and over time for any individual, and so there seems to be limits to our ability to synthesize emergent properties from network topologies using standard approaches. There is a famous example more than 150 years old where the emergent properties of a very-large complex adaptive system were discovered by an investigator who relied only on stimulus-response learning methods.

3.5   Knowledge Discovery by Non-reductionist Analyses

The most successful of all non-reductionist analyses in science led to the birth of a field known as evolutionary biology. Charles Darwin's discovery of how evolutionary "forces" came to shape the diversity of life on Earth [Darwin (1859)] was chronicled during his five-year journey as a geologist and naturalist aboard the HMS Beagle.

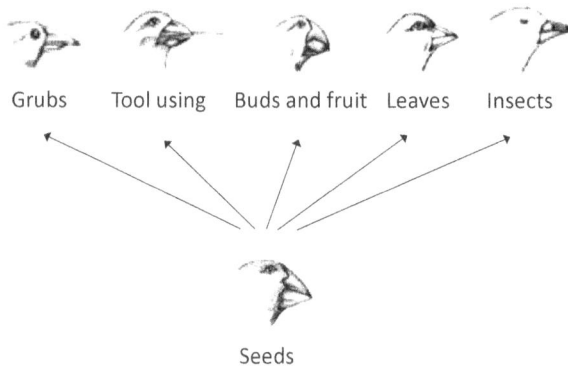

Fig. 3.6   Darwin's finches illustrate adaptive radiation where one species diversifies into many to exploit the habitat.

Darwin visited the Galapagos Islands, which are about 600 miles west of the South American continental coast. On the islands he found many more sub-species of birds than he did on the mainland coast. He compared the two ecosystems to find explanations for differences in bird features in what ostensibly appeared to be the same species of bird. Darwin did not reduce each ecosystem to its basic elements to gain knowledge; instead he studied ecosystems as whole entities and at different spatial scales, relating stimulus with response in order to discover governing principles.

One of the things Darwin discovered was that the proliferation of some finch subspecies on an island was related to their greater ability and need to take advantage of food sources on that island. The feature he selected was variability in the shape of finch beaks (Fig. 3.6). Over time, those birds with beaks that best matched the available food sources proliferated more than others. Because beak features are inheritable, natural selection led to a greater number of that subspecies on the island. The sequence of any anatomical feature change from a distant common ancestor through present time can be verified through fossil records.

Darwin hypothesized that *speciation* was due to selection of the *phenotype* (characteristics that result from an interaction between genes and environment) most able to meet an environmental challenge from the options provided by natural variability in the population. What is remarkable is that he did this at a time before genetic concepts were established. His investigations have been repeated since that time for all types of animals and plants to show that natural selection is a unifying principle of life on Earth.

Eigen-analysis should be appropriate for studying evolutionary biology. The inputs of interest to this local region within the bird system are food sources, and the outputs are beak shape and proliferation rate. The system Darwin described is always in dynamic equilibrium with sufficient genetic variability to say individual birds are unique. Variable beak shape results from different eigenfunctions (via genes) at play within the finch population. The system in this case is quite complicated because it encompasses the social structure of area birds within their environment and the finch genome, which is the history of the species along its branch of the phylogenetic tree all the way back to the origin of life. By narrowing the range of food sources to mostly seeds, for example, the system responds to this input by resonating with specific phenotypic modes within the bird system. That resonance amplifies the phenotypic output of a short thick beak, as illustrated in Figure 3.6, by increasing the relative proliferation rate of that subspecies compared to others.

In the guitar-string example, by placing fingers on the fretboard, the guitarist determines which eigenfunctions can be activated from the many it is capable of generating. In the finch example, restricting the food source preferentially selects the eigenfunctions that manifest a specific beak shape, which effectively increases proliferation for that subspecies relative to others. We do not know how to compute the eigenfunctions of the finch system like we can for the guitar string, but the two complex systems share features that make an analogy reasonable. The bird system, in Darwin's view, uses natural genetic variability among breeding individuals to provide options in dealing with environmental inputs. Once that input is applied and the appropriate eigenfunctions are activated (genes are expressed), that sub-population of birds increases their numbers over others. Nevertheless, the bird system remains in equilibrium even as it adapts to those environmental inputs. There doesn't seem to be any transitions in the Darwinian approach, so we are likely to be missing parts of the story if, as some contend, punctuated equilibrium is a more accurate evolutionary model of life on Earth.

We have only touched upon one tiny area in the extensive field of evolutionary biology, and yet we believe that eigen-analysis could serve as a framework for other studies in that field.

Although non-reductionist analysis appears capable of explaining the emergent property of natural selection, the reductionist approach of physical sciences also has an important role in furthering our understanding of the mechanisms of evolutionary biology. Reductionism can explain the molecular processes underlying these actions. Mechanisms describe how the cellular machinery engages the large-scale stimulus (food sources leading to blood-born molecules) to facilitate the evolutionary process; they help us to understand how cells within individuals can help populations adapt to their environment. Although adaptation may not be directly encoded in the molecules of cells except to provide inheritable phenotypic variability that will interact with the environment to produce that large-scale emergent property, the connection must be important. For example, if we knew which cellular-scale features to monitor so that we could track changes in population-scale eigenvalues, it might be possible to observe evolution in action. Experiments could be performed in species like drosophila (fruit flies), a standard animal model of evolutionary biologist because of their short generational cycle.

In summary, the history of science and engineering offers many examples where reductionist and non-reductionist approaches were combined to successfully describe the behavior of large complex systems. We believe continuing the combined approach is the proper path forward in the discovery of knowledge about complex systems, and that eigen-analysis can serve as a framework for the non-reductionist approach.

3.6   Known, Unknown and Unknowable

As we authors began to explore various approaches to knowledge discovery involving complex systems, we kept bumping into questions about the limits on our abilities to acquire and process system information. Some of these thoughts are included in this section.

Knowledge is itself a complex adaptive system. It exists in a rapidly-changing dynamic equilibrium state where its components are a variable distribution of social minds that regularly exchange information among themselves and their environment. In the past, knowledgeable people were those who retained the most information when exchanging with others because knowledge depends on having fast access to relevant information. Today, instant access to vast information repositories is now available to almost anyone anywhere. So, as it is frequently said, knowledgeable people of the 21st century are those most able to quickly extract meaning and value from information archives, which of course is how new knowledge is developed. The skill set valued by the new information-based economy is challenging educational programs to adapt to these changing societal needs.

What is *known* to the world today is the value of communicated information that resides within each of us and that which was processed by others and recorded for us in accessible form. For example, we know how cells receive signals from the body and use DNA to respond. What is *unknown* today is the full value of the information we now possess or could obtain about systems we only partially understand and hence cannot yet formulate appropriate questions. For example, we don't yet know how DNA is involved with large-scale emergent properties like human behavior. *Unknowable* matters are those about which information cannot be obtained. For example, we speculate that we cannot be certain how DNA molecules first came to exist. There are several inherently unknowable domains that have already been proven to be so. We now briefly discuss two cases.

The first case is in the mathematical field of number theory and logic. This field is summarized in a three-volume book "Principia Mathematica" by Alfred North Whitehead and Bertrand Russell [Whitehead and Russell (1927)]. The book series was quickly recognized as a landmark in mathematical philosophy for its presentation of attempts to arrive at a set of axiomatic rules that were to become foundational. Reductionist heaven! *Axioms* are starting points of a reasoned argument that are widely accepted as true without additional proof. Whitehead and Russell ambitiously attempted to present a set of axioms in logic that could be used to prove any mathematical fact. If successful, then those axioms would essentially contain all of mathematics in some form, and they would have shown there are no limits to knowledge discovery through mathematics. This is powerful stuff, on par with the Theory of Everything from physics discussed earlier.

In 1931, Kurt Gödel threw a monkey wrench into such efforts when he presented the *incompleteness theorem* [Gödel (1962)] proving their overall goal was not achievable. He proved that for any set of axioms in logic, there will always be some mathematical fact that will fall outside that set and hence cannot be proven by the set. This is a clear example of a limit to knowledge discovery in mathematics; the sought-after set of logical axioms that encapsulates all mathematic is unknowable.

The second case illustrating the limits of knowledge involves randomness. Chapter 6 is devoted to the more general topic of variability in systems because of its great importance to the formation of complexity. Here we discuss only one aspect. If everything about a system was deterministic at all levels and at all times, then it might be possible to gain complete knowledge of that system. A *deterministic system* is one that generates exactly the same response each time a fixed stimulus is applied if we start from the same initial conditions. A *variable system* is one that is not strictly deterministic and yet generates responses that convey some amount of information. A *random system* generates variable responses to a fixed stimulus with no discernible information about the input. All real systems are variable, although specific cases can range in their response toward either the deterministic or random extremes.

Beginning with Kepler and Galileo in the 16th century, classical physics was developed to precisely describe the macroscopic world we knew about at the time. These laws are based on determinism, where all properties of system components are known exactly. At the beginning of the 20th century, however, quantum physics was developed to describe the sub-nanoscopic world on the scale of atoms and smaller. Quantum investigations taught us that systems operating at this tiny scale have some properties that cannot be known exactly at the same time; e.g., the exact simultaneous position and momentum of an electron are unknowable. Our inability to know this information is not because we don't know how to make accurate measurements; it means that nature prevents us from knowing both properties exactly at the same time. You might imagine the controversies these ideas caused when first introduced about a century ago. A major consequence of quantum theory is that there are fundamental limits to knowledge. This humbling finding was an enormous shock to even the most prominent scientists of the day, just as Gödel's theorem was a disappointing surprise to mathematicians at about the same historical time. There are clearly limits to the reductionist search for the most fundamental components of the universe. So, system explorers must always remember to ask about what is known, unknown, and unknowable.

Regarding complex systems as a general concept, we would venture the following. What is known about any and all complex systems can be summarized in a set of universal principles, perhaps involving the four propositions listed in Section 2.9. We also suggest that new knowledge can always be obtained about a specific system from stimulus-response learning.

What is unknown about complex systems is how to find key observable summary features – system indicators – that can reveal changes in eigenstates that characterize emergent properties and their all-important evolutionary trajectories. As mentioned in Section 1.3, professional people (we used examples of economists, farmers and physicians) routinely locate key indicators of their systems to predict important changes that are about to occur, even though they do not have complete knowledge of each system they are investigating. This can be done, we have seen it, and yet we don't yet have a systematic way of implementing a search for reliable indicators. Identifying reliable behavioral indicators for a system is among the grandest challenges of the 21st century engineering.

What are unknowable about large complex adaptive systems, such as living organisms and economies, are complete structural mappings of individuals that allow us to conceptually reconstruct the system in order to understand and accurately predict future behavior. These systems exist in dynamic equilibrium states that continuously change and adapt over time such that the topological maps are constantly in flux. The resulting key question is how to acquire indicator-class information about these systems that permits characterization and prediction of future behavior. We submit that identifying the currently-unknown indicators, which makes system properties known, is a way around the unknowable roadblock of having a complete mapping of a system topology. And, of course, we feel that eigenanalysis can play a central role in the eventual solution.

Variability is clearly essential for biological evolution, as Darwin found, and we believe it is an essential element of all complex systems. Variability can manifest itself as redundancy and multi-functionality in organisms, where it provides robustness to internal noise and component failures. For example, a baseball club may have two second basemen on a team or one player who is competitive at multiple positions. When a starter pulls up lame, redundancy and multi-functionality enable teams to achieve their objective of winning a baseball game even if the team behaves a little differently while doing so. In the extreme, consider the biological systems of each player. Billions of cells out of the trillions in a human body die each day. And yet we maintain functionality and are robust to all sorts of environmental challenges, ultimately because of the influence that variability exerts on the formation of an equilibrium state.

3.7   Preparation for Knowledge Discovery

Science historian Charles C. Gillespie once remarked that "the first trait of a profession is the custody and development of a body of knowledge and the second is the provision of economically viable careers." [Gillespie (1983)] Indeed, there is employment within all topical areas of business, government, and other fields that focus on knowledge discovery, and there are business models that sustain these professions. The business model for knowledge discovery itself has a strong influence on direction and rate of discovery, and there is the important associated matter of how to train the next generation to undertake research in highly integrative and rapidly evolving areas. Consequently, we briefly review how research is funded in the USA today, and how this business model strongly influences education and research.

The business model that maintains the equilibrium state of our complex knowledge-discovery systems (universities) began during World War II. American universities had been multiplying as a result of the sequence of Morrill Land-Grant Acts passed following the American Civil War. They provided a means for developing the rich network of land-grant colleges that were modeled after hugely-successful European universities. These American institutions had grown during the 80 years between the Civil War and WWII. Just before WWII, engineering research and education at American universities focused on agricultural sciences, medicine, and civil, mechanical, electrical, mining and metallurgical engineering. These efforts were primarily funded by state governments hoping to train a workforce for farming and industry. They were also funded by private foundations, donors, and student tuition. Federal funding for science research was small, supporting mainly the Smithsonian Institution, the U.S. Geological Survey, and agricultural experimental stations. The role that the federal government assumed for science and technology that began during WWII fundamentally changed forever the relationship between scientific research and the US federal government. For a thoughtful and fascinating historical summary, see [Stokes (1997)].

In 1941, President Franklin Roosevelt named Vannevar Bush to be director of the Office of Scientific Research and Development. Based on Bush's long career in academics, industry, and administration (e.g., Electrical Engineering Professor and Dean at MIT, president of the Carnegie Institution of Washington, founder of Raytheon Corp.), he was well prepared to become the first US Presidential Science Advisor [Zachary (1997)]. He would organize nationally-coordinated wartime efforts that notably accelerated such disruptive technologies as atomic weapons, analog computers and radar (acronym for radio detection and ranging). During the war, Bush championed the idea of the federal government fully funding all of the research on projects of national interest that took place at universities and in industry, including a schedule for paying overhead costs. That decision was controversial at the time because the policies for conducting classified research outside government labs and assigning intellectual property was yet to be established, and yet the eventual results spoke for themselves.

Near the end of WWII, President Roosevelt asked Bush to draft a proposal that outlined a continuing US-government role in post-war research. Experience showed that government funding could fuel a public economic engine by accelerating developments in fundamental research that would lead to new technologies and thriving new businesses. His vision was ultimately realized through the last half of the 20th century. Bush prepared a report in 1945 entitled *Science – The Endless Frontier: A Report to the President on a Program for Postwar Scientific Research* (later published as [Bush (1960)]) that outlined a plan for federal funding of basic science research at elite universities that was based on wartime experiences.

After five years of work, more reports, and modifications following much political haggling, the US Congress authorized funding for the National Science Foundation (NSF) in 1950 to support investigator-initiated basic-science research conducted at qualified institutions in a way that was outside the direct management of the federal government. During the 1950s, this effort snowballed into several new federally-funding agencies, most notably the Office of Naval Research (ONR), the Atomic Energy Commission (AEC), Advanced Research Projects Agency (ARPA, now DARPA; D standing for Defense), and the extramural program from the National Institutes of Health (NIH). In 2015, the federal government awarded about $70 billion to non-defense "science and medical research," which was less than 2% of a $4 trillion federal budget.

Of course, corporations also fund manufacturing- and service-sector developments. In fact, the Battelle Institute estimated in 2014 that $465 billion (2.8% of US GDP) would be invested by US industry R&D and $1.4 trillion at corporations worldwide in 2014, but this research is not investigator initiated (curiosity driven). Perhaps the most public success of corporate research investment into basic science occurred at Bell Laboratories, a research and development division of AT&T Corporation. A highly readable history of the broad spectrum of scientific and technological achievements at Bell Labs from 1920s until the 1980s when the government broke up AT&T can be found in [Gertner (2012)].

Today, federal agencies fund most of the curiosity-driven research in science and engineering. The role of academic engineering is to (1) conduct and disseminate research that shows the world what is possible when fundamental knowledge is applied in the act of solving important problems. It is often the first step in translating knowledge into goods and services. And (2) prepare a workforce capable of solving the problems of tomorrow. The funding model that supports academic engineering is critical for facilitating both aspects. Indeed, scientists and engineers with advanced degrees make up the workforce at the funding agencies, who then recruit practicing scientists and engineers throughout academia and industry to formulate special requests for proposals based on national needs and to evaluate all proposals submitted to these agencies. This highly-egalitarian peer-review process is well regarded throughout the world and emulated by other countries because of its integrity and effectiveness at selecting the best ideas for investment by the nation.

Since the 2008 recession, US federal government appropriations have lagged behind university needs at the same time states have reduced their budgets to public universities and colleges. Many who depend on federal funds feel that the pressure exerted by lower congressional appropriations has caused peer reviewers to be more conservative in rating proposals for funding. Scientific reviewers are taxpayers like everyone else; they too want to be sure the investments payoff in new goods and services. Taking their roles as protectors of the public trust very seriously, reviewers understandably tend to value proposals from large established research groups with long records of working together on important problems as the best investments. While this approach moves forward projects steeped in current thinking and needs, it tends to suppress more risky ventures and thinking outside the norm that might be required to advance very large-scale and difficult problems involving systems.

We often hear of new efforts by federal funding agencies to enhance the translation of research findings into everyday life (sometimes called technology transfer), either through new goods and services or by building new skills in the next generation of scientist and engineer. It may not be widely known among the public that a large fraction of research funding goes to supporting graduate student research and education. It is true that the system is not as efficient at technology transfer as it could be, and so less economic development results from these knowledge discovery investments. So the trend today is to give preference to funding projects that have a clear translational path. While this plan enhances translation, it can suppress new curiosity-driven ideas that don't yet have a clear path toward translation especially in tough economic times. We will give two examples of curiosity-driven research that had no translational path at the time ideas were developed but later had a major influence on progress.

Before the 20th century, matrix mathematics was largely an intellectual exercise as a tool for solving many equations at the same time (systems of simultaneous differential equations). Then in the 1920s, physicists like Heisenberg, Born, and Jordan who were searching for a way to represent discrete-energy transition probabilities in atoms as quantum mechanics was being developed, were able to adopt this mathematical tool and make quick progress.

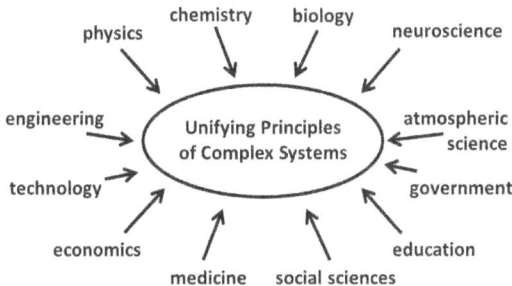

Fig. 3.7 Shown is a unified, multidisciplinary diagram for building understanding of the behavior of complex systems. The approach aims to apply unifying principles to problems from different disciplines. Just a few of the areas are indicated. Understanding complex systems is a common goal of many fields of study.

Also in the 1970s, biologists Howard Temin and David Baltimore each led groups that discovered the enzyme known as reverse transcriptase that enables retroviruses to inject their messenger RNA into cells that can integrate into the host DNA to propagate. Their immediate application of this information was to explain if it were possible for viruses to initiate cancers. Yet this body of work was available to pioneers like Robert Gallo who began studying infections of another retrovirus, HIV, in the 1980s. AIDS research was sped up because of earlier basic-science work on retroviruses.

It is very hard to predict which ideas will ultimately translate into useful tools, much less new goods and services, and so funding agencies and the entire academic community continue to wrestle with these issues as we all want to maximize the return on public investments. Stewarding support of knowledge discovery with public funds can be both very challenging and very rewarding.

There will always be debate on the directions and levels of public investment in research. Debate is exactly what is needed for progress. To help taxpayers make the best decisions, scientist and engineers need to do a better job at communicating to the general public the needs and potential returns on research investments. We need to keep in mind that the knowledge system on which we all depend to make decisions is in a dynamic equilibrium state, and that research not only generates new knowledge but it speeds the circulation of existing knowledge through the network that strengthens the system. We may not need to change the funding model for knowledge discovery to accelerate advancements because it is already well designed for this activity, but we need to have discussions like the ones we are hoping to precipitate with this book to raise awareness that the complex systems that surround us have unifying principles. To further our knowledge about system behavior, we need to have experts from many fields of knowledge (Fig. 3.7) join together with other disciplines to build upon and test principles that are common throughout all disciplines.

Furthermore, we need to migrate away from discipline-specific student training and toward problem-based training. But we need to do this carefully. The migration has been happening at top research universities, like the University of Illinois at Urbana-Champaign, since the 1970s. Increasingly more research is conducted at problem-focused institutes, centers and laboratories on university property than in the past where almost all research was conducted in discipline-based departments. We need methods that overcome the disciplinary language barriers that we each have developed, and we need to do this without compromising the quality and integrity of the research and training. Yes, education is also a complex adaptive system. Funding agencies already play an important role in graduate training focused on the broad interdisciplinary topics of grand-challenge-scale problems. We need to continue to push universities to innovate in these areas and reward those who can achieve the broad training needed for solving problems in the 21st century.

Chapter 4

Examples of Mechanical Systems Undergoing Transitions

4.1 Introduction

Our goal thus far has been to illustrate for readers the ubiquitous influence of complex systems in our daily life. Some believe complex adaptive systems are a rare special case. We find just the opposite; most (maybe all) systems found in nature express some degree of self regulation, emergence, and readily-observed nonlinear properties that can change as that system responds to input stimuli. As we argued in Section 1.6, by this definition, objects as inert as rocks and wooden rods can be considered as complex systems. Since so many systems are both complex and adaptive, efforts made to further understanding the general principles regulating such systems will have a broad impact on science, engineering, and society in general.

We suggested that a generalized version of eigen-analysis would be useful for discovering the uncoupled properties of these systems, and for predicting property changes such as a slow evolution of the equilibrium state and a sudden transition to a new state. Because observable system properties can be described as combinations of eigenstates that define the state of equilibrium for the system, eigen-analysis provides a natural basis for developing models that predict the responses of all types of systems to input stimuli, from inert rocks to populations of multicellular organisms.

Our discussion has been intentionally descriptive and nonmathematical for two reasons, despite the highly mathematical nature of these topics as described in the scientific literature. First, we want to engage as many thoughtful readers as possible regardless of background because we believe the importance of eigen-analysis extends well beyond the world of traditional science and engineering. Second, we believe reductionist approaches based entirely on mathematical models are limited (a) in their ability to reveal all emergent system properties at every scale and (b) by the fluid nature of the adaptive component topologies of complex systems, which makes it difficult to completely map all system components as required for mathematical analysis. Despite limitations, reductionist approaches are extremely valuable for developing models that describe functional mechanisms. Many believe that ultimately mathematical models will be effectively merged with data-driven machine-learning methods to find better solutions to engineering problems. This hybrid approach merges the accumulated experiences of stimulus-response data with mathematical models to predict the behavior of systems too complicated or rapidly evolving to model using only mathematical methods. We hope this chapter provides readers with templates for considering familiar experience from a systems perspective.

Here we discuss nonmathematical examples of mechanical systems that are described mathematically throughout the engineering literature. Our goal is to build reader intuition through graphical explanations. With each example, we hope to reveal the underlying principles at work. We begin with simple mechanical structures, support columns and arches, before ramping up the complexity to discuss granular materials and the movements of tectonic plates. While we cannot yet offer detailed procedures for discovering and monitoring the eigenstates of large (many component) systems, we believe the best way to illuminate analysis paths forward is through analogy with graphical descriptions.

4.2 Eigen-analysis

Before presenting examples, we briefly summarize standard eigen-analysis of linear systems and outline the extensions needed to study large, nonlinear systems.

Traditional eigen-analysis is already familiar to scientists and engineers as a way to mathematically represent *uncoupled properties* of a *linear system* operating at or near *equilibrium*. Eigen-analysis is used to both separate and reassemble system components into primary elements. This is possible for any system that can be expressed numerically, which is a principal reason for modeling (Ghaboussi and Wu, 2016). Eigen-analysis can be a tool for both reductionist and non-reductionist investigations. However the results of eigenanalysis can be challenging to interpret physically. Analysis and interpretation through any functional decomposition of a model is an engineering fine art.

Eigen*vectors* [1] describe connection patterns or pathways through highly-interconnected networked components that enable an emergent property of that system to be expressed independent of other properties. This is what is meant by *uncoupled properties*. The eigen*value* associated with an eigenvector describes combinations of material and geometric features of system components that weight the importance of that eigenvector in expressing a specific property. Remember, system properties are what determine responses to an input stimulus, and so the intuitive connection between properties and eigenstates is fundamental to predicting system behavior. We gave examples of eigen-analysis in terms of the properties of wooden rods, guitars, and suspension bridges in previous chapters as a way of showing how engineers can design desirable properties into the devices we build.

In static systems, like the wooden rods supporting someone sitting in a chair, eigenvalues are real numbers. In dynamic systems, like a guitar string vibrating at natural frequencies, the eigenvalues can be complex numbers. Importantly, eigenvalues of linear systems operating near equilibrium are constants that do not change with time. Therefore, properties of linear systems operating near equilibrium do not change over time.

Nonlinear systems have eigenvalues too, but they are not constants. These eigenvalues can be functions of space and time that reflect environmental conditions and the applied input. For example, your eyes and ears adapt nonlinearly by changing their sensitivity as the ambient light and sound intensities vary when you walk into a dark, quiet house near traffic on a sunny day.

Formulating and interpreting mathematical models of very large, nonlinear systems can be difficult, which discourages use of traditional eigen-analysis in describing them; that is, unless a system is being operated in a state close to equilibrium where their behavior is approximately linear. When we mention this approach to well-trained scientists and engineers, they immediately say, "Wait a minute! Eigen-analysis is for analyzing linear systems!" However, they also realize that nonlinear systems can be linearized continuously over time, and so clearly the concept generalizes.

We need flexible models of the systems we hope to study that predict behavior for a range of responses if we are to have a prayer of solving the big

---

[1] The eigenfunctions of continuous systems, like guitar strings, were discussed in previous chapters. Eigenvectors apply to systems with discrete components, like grains of sand in granular materials or cells in the body. For descriptive purposes, we treat both terms as equivalent and use them interchangeably.

problems involving systems discussed in Chapter 1. You see, all systems are inherently nonlinear. Linearity can approximate the input-output responses in some systems over limited parts of the response range, but those models may not be able to predict the extreme behaviors of greatest interest – *transitions*. Eigenstates can help us predict the behavior of all systems once we learn to measure them for systems too complex for mathematical analysis.

A very interesting thing about nonlinear systems is that their eigenvalues change in ways that predict future behavior of the system. For example, when the real part of any initially-nonzero eigenvalue approaches zero, the system begins a sudden transition to a new equilibrium state. Even the largest and most complex systems can be monitored this way if we can find ways to track their eigenvalues.

4.3   Choosing a System Model

Modeling is an important step toward understanding systems. One way that a system modeler can represent a system of discrete components that is responding to environmental stimuli, at least approximately, is through a matrix of values. For example, a transformation matrix is a grid of numbers that explains how an input vector of numbers is processed by the system matrix to generate a vector of output numbers.[2] Using the rules of matrix algebra (taught in high school these days), a system matrix can be multiplied by a vector (a long vertical string of numbers or functions) representing input stimuli, and the result is a new vector that describes the output of the system as it responds to that specific input.

A system matrix describes system properties by virtue of its ability to map applied stimuli into observable responses. Hence eigen-analysis of a known system matrix yields its uncoupled properties in the form of

---

[2]If you are math shy, please avoid reading the following. Output vector **y** is modeled using the product of matrix **A** and input vector **x**. The vectors (bold lower-case letters) and matrix (bold upper-case letter) describe a model via the shorthand equation **y** = **Ax**. Written out in detail the same equation appears as,

$$\begin{pmatrix} 13 \\ 14 \\ 2 \end{pmatrix} = \begin{pmatrix} 1\ 0\ 2\ 5 \\ 2\ 4\ 1\ 3 \\ 0\ 1\ 0\ 1 \end{pmatrix} \begin{pmatrix} 2 \\ 1 \\ 3 \\ 1 \end{pmatrix} .$$

This is a linear equation that describes a simple linear system with four input values and three output values. It is a linear equation because the only operations needed are addition and multiplication.

eigenstates. The all-important interpretation step is assigning meaning to each of the numbers or functions within the matrix and the associated vectors. This is just one of several common mathematical representations of a system in operation. If we are good at reducing a system to a complete list of its component elements, that system matrix is said to be "known," and it can be straightforward to decompose its matrix, even a large one, into eigenvalues and eigenvectors on a laptop computer in a few minutes. The more we understand about a system's internal structure and function, the more accurate and valuable these mathematical representations become. A model's usefulness is evaluated by comparing the accuracy of its predictions with experimental observations.

Since there are many ways to describe properties of a system, there are just as many legitimate mathematical models of a system and each has its own set of eigenvalues and eigenvectors. An eigenvalue to an engineer is like a wrench to a mechanic; it has little intrinsic value. Value is revealed only when it is used to solve a problem. If you have worked on a car, you quickly realize there are many different wrenches that can be used to remove a bolt. Some are more useful than others depending on the accessibility of the bolt. As with mechanics using their tools, modelers using eigen-analysis require deep knowledge of the overall system, sharp skills using mathematical tools, and sometimes a little luck and a few band aids. At best, a model is an accurate approximation of some limited aspect of a real system; or as statistician George Box famously wrote, "essentially, all models are wrong, but some are useful" [Box and Draper (1987)].

Linear models are of little use in predicting sudden equilibrium-state transitions – a distinctly nonlinear response of systems to stimuli. Transitions occur when at least one of the nonzero system eigenvalues reduces to zero. The utility of a model depends on the questions being asked.

To illustrate, consider the black curve in Figure 4.1 that is found by measuring how stimulus s generates response r when it is applied to some unspecified but simple system. Each point on the black line describes a state of the system in terms of this observable response. A stimulus can be light photons incident on a sensor and the response can be a voltage generated at the sensor's output. Or a stimulus can be a force applied to an object and the response can be the displacement caused by that force.

The example system in Figure 4.1 has a single dimension, meaning there is only one stimulus-response pair. There is only one dimension; i.e., only

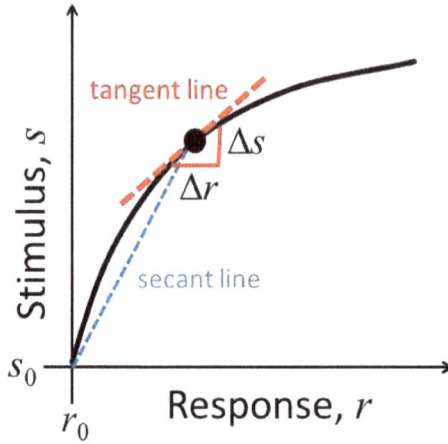

Fig. 4.1   Total (dotted line) and incremental (dashed line) representations are illustrated for the stimulus-response curve given by the solid black line. This system's properties are described using two models given, respectively, as a set of secant-lines and a set of tangent-lines.

one response of the system to input. Hence this phase plot[3] is a line. Two-dimensional systems are represented by planes in phase space. It is important to point out that practical systems typically have many dimensions, sometimes thousands, which are represented as hyperplanes in a high-dimensional phase space. High-dimensional spaces are very hard to visualize until we project onto lower dimensional subspaces. To avoid that complication we limit this discussion to a 1-D system with knowledge it can be generalized to higher-dimension systems. This topic is discussed in greater depth in Chapter 6.

Since the stimulus-response curve is not a straight line, we have clear evidence of nonlinear behavior in this response range. A common mathematical modeling approach is to *linearize* the system, which is to represent its behavior from a set of straight lines. One technique is developed by asking: What is the response of the system to a small (incremental) stimulus applied right now? The incremental or *tangent-line properties* are found by adding a tiny stimulus Δs to the current value s to apply $s + \Delta s$. Then we measure $r + \Delta r$, which is the tiny incremental response Δr that appears

---

[3]For those with some math background, *phase space* refers to a vector space in which it is possible to find observable states of the system. A *phase plot* defines specific system states in phase space for a set of experimental conditions. These plots generally eliminate independent variables like time and position.

in addition to the current response r. These incremental properties are quantified simply by the slope, $\Delta s/\Delta r$, observed at each reference point in the (r, s) plane. Clearly the stimulus-response curve in Figure 4.1 has tangent properties that change with r; the slope is large at small values of r and small at large values of r. The eigenvalues are given by $\Delta s/\Delta r$ for this simple nonlinear system, which is always in equilibrium within the range shown even though the eigenvalue changes because small Δs always produces small Δr. Note that the single property for this system (quantified by the slope of the curve) is not constant, rather it is a function of response r. This behavior defines a nonlinear system. A system that is linear over some response range has constant properties over that range.

We might ask a different question: What is the response of the system to a stimulus applied beginning sometime in the past. For example, let's select a fixed reference point at the origin of the plot (r_0, s_0) and draw a line from the reference point to each point on the plot (e.g., the dotted line in Figure 4.1). In that case, we find the net or *secant-line properties*. These can also represent the system by a set of straight lines with slopes that change with r but not in the same way as the tangent-line properties. The incremental and net properties are equal only for linear systems, where properties are constant over the range of stimulus values appropriate to the question being asked of the model. Both models employ a set of straight lines to represent the observable data (r, s), but each model describes different properties. Both models are "correct" in some sense but which model is more useful?

The distinction between these two system representations that are both based on stimulus-response data is with respect to the reference point. We fixed the reference at the origin for the secant-line properties in this example and allowed it to vary for the tangent-line properties. The reference point selected may not seem important but it is critically important for interpreting system properties. A system that changes in response to a stimulus is a slightly different system at every observable state. If we want to know how a system changes in response to an input we must define change relative to a reference at an earlier time, and there are an infinite number of earlier times to consider. Selecting a reference time introduces modeler bias into the representation of the system, so we must choose carefully.

One reference point is more privileged from the model utility perspective than all others. It is the reference used by the tangent-line or incremental-property representation. Because nonlinear systems can change properties depending on their state, the current state is the most important reference point for answering questions related to predicting behavior in the next instant. If we don't know the current state of the system they must be measured. Our knowledge about that system will be limited because every measurement contains errors. There are occasions where it is advantageous to use secant-line properties for modeling, e.g., for measurement noise suppression. In reality, however, system properties exist regardless of how well we know them, and the instantaneous properties, where eigen-analysis holds, are the only ones for which transitions occur when eigenvalues go to zero. We will stick with the tangent-line models in what follows. Now we are ready for examples.

4.4   Equilibrium-state Transition: Ideal Support Column

Consider first the example shown in Figure 4.2. The stimulus is a force applied downward along the axis of an ideal support column, and the response of interest is the position of the point on its upper free end. The column is ideal in the sense it is perfectly straight, radially symmetric, and made of a material with perfectly-uniform mechanical properties. As force is applied, this ideal column is compressed a small amount. If we then steadily increase the load, at some point the column will suddenly buckle and collapse.

We are examining the movement of just *one point* on the free end of this two-dimensional column in the plane (u, v). There are two dimensions of motion as indicated by positional axes u (horizontal) and v (vertical). This geometrical description suggests two eigenvectors, e.g., the standard basis $(1, 0)$ and $(0, 1)$, and two corresponding eigenvalues. The first eigenvector $(1, 0)$ describes horizontal motion of the free end along the u axis while the second $(0, 1)$ describes vertical motion along the v axis. Weighted combinations of these two eigenvectors describe all possible movements of that one column point in the plane. If we adopted a three-dimensional model, we would need three axes (u, v, w) and three eigenvectors, e.g., $(1, 0, 0), (0, 1, 0), (0, 0, 1)$. For simplicity, let's stick with a 2-D model. The 2-D column has a pair of displacements for each point within the column, and so there are actually an infinite number of eigenvalues and eigenvectors in the column. We focus on just one point at the tip of the free end to simplify the description.

The eigenvalues associated with our eigenvectors depend on the shape of the cross section of the column, the initial stress on the column, and the material properties. Their numerical values are unimportant for this discussion except to say both are positive real numbers. Large eigenvalues

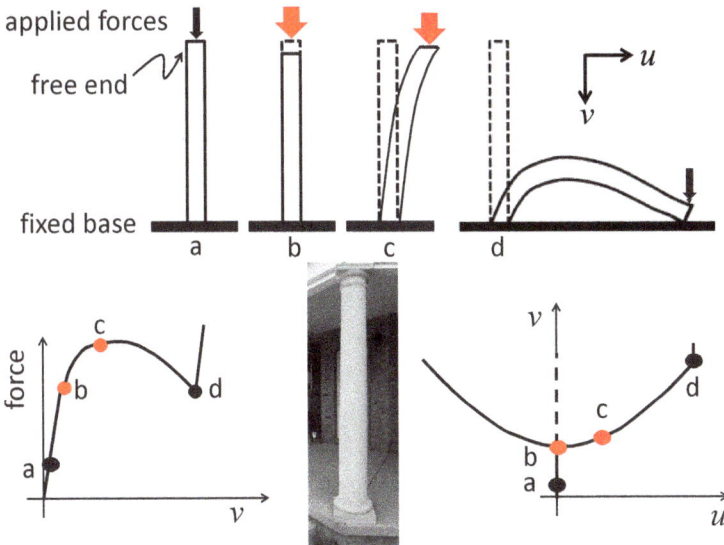

Fig. 4.2 Downward vertical forces are applied to the upper free end of a column along the vertical axis labelled v. We measure column properties from the response v as a function of the stimulus force (lower left plot). At time-point (a), a small force (small force arrow at column top in a) is applied that compresses the column a small amount along only the v axis. Increasing the force beyond its linear-elastic range (large force arrow in b) causes the column to initiate a buckling transition as an eigenvalue goes to zero. Maintaining the force, the column now begins to move along both the u and the v axes (c), and column collapse in inevitable. Even if the force is reduced (small force arrow in d) the column is now destined to collapse. As it comes to rest, any additional force will further compress the column material. This time course of events can be followed in the two phase plots in the bottom row. On the lower right, we plot the 2-D position of the free end in the (u, v) plane at the same four experimental times. The v axes in both plots are the same.

tell us the column is able to resist movements when even large forces (loads) are applied.

The column might be designed to support loads as a linear system. If one's office is in a building held up by this support column, interest in the nonlinear properties engaged at high loads outside the designed operating range would increase significantly. Each column is designed to support vertical loads with as little movement of the free end as possible. So when a small vertical load is applied, as in situation (a) in Figure 4.2, the top of the column is compressed downward a small amount in the direction of the second eigenvector $(0, 1)$ and not at all along the first $(1, 0)$. As long as the stimulus-response curve is linear, an applied force in the direction of $(0,1)$ generates no movement along $(1,0)$, as we say these properties are uncoupled. The column is also behaving elastically because, if we were to remove the load, this ideal column would spring back to its original height. This situation describes classic linear-elastic mechanical behavior.

Slowly increasing the downward load over time, we see that between points (a) and (b) the movement response remains linear with force and entirely vertical as the column is compressed. As long as we do not exceed the buckling load, the system is linearly elastic and operating near its static equilibrium position, $(u = 0, v = 0)$. The linear model is predicting the behavior being measured. Great!

What happens if we add loads beyond the buckling load? To examine this question, we plot on the lower-right side of Figure 4.2 all movements within the (u, v) plane. The dots on the curves show equivalent points on the curve of force versus vertical displacement to its left, and shade of the dot corresponds to that of the arrow at the top of the column indicating the amount of applied force.

Increasing the load beyond that shown in situation (b), the force-displacement plot for the column responds nonlinearly; that is, the displacements are no longer proportional to force and the eigenvalues are no longer constants but functions of the applied force. Hence the column that was at one time in a static equilibrium state is now in a rapidly evolving equilibrium state with properties that are changing with time. In columns made from stone or steel, the material properties reflected by the eigenvalues would likely change (stone crumbling and steel yielding), but in the ideal-column example of our imagination we assume only the geometric properties are changing. The linearly-elastic properties of the material remain unchanged during the experiment. Although the column folds over as it buckles, it will spring back once the load is removed. Hopefully you see how models simplify reality so that we can focus on specific features and ignore others.

At point (b), the eigenvalue of the first (horizontal) eigenvector in this example approaches zero, which begins the buckling transition that eventually ends in a new static equilibrium state. At the zero-eigenvalue moment, it takes essentially no additional energy to move the free end of the column in the direction of the first eigenvector $(1,0)$. The top end slips horizontally toward the right but it could have gone left or even continued compressing downward. Immediately after the transition at (b) begins, neither eigenvalue is zero anymore. Because zero-energy motion is no longer available, column motion becomes a combination of both eigenvectors and movement commences along both axes. The transition continues with the free end of the bending column moving both vertically and horizontally now, as shown at point (c) in Figure 4.2, until it settles at the base at point (d). At (d), the transition has ended and a new equilibrium state is established. Increasing the load further only compresses the elastic column material, but removing the load altogether returns the column to its original height along the same motion path.

The zero eigenvalue in this example was caused entirely by geometric nonlinearities because our ideal-column material is assumed to be perfectly linearly elastic. That's how it is able to bend reversibly without material damage. Any amount of material nonlinearity and this transition will not be entirely reversible. By interpreting the eigenstates in this manner, we obtain significant information about nonlinear system properties even though we are only monitoring the movement of one point at the free end. If the situation becomes more complicated, as occurs with material and geometric heterogeneity, then we might need to wisely select several more points in the column to monitor and analyze to accurately predict behavior under general conditions.

Before buckling there was only one possible solution path for the column end – to move downward. Force was applied exactly along eigenvector $(0, 1)$, so the linear response of the column could only be along the v axis. At the instant when the eigenvalue equals zero, suddenly three solution paths open for motion of this ideal column. Continuing along the vertical path is possible in this ideal column because that is the direction of the applied force. More realistically, the column is much more likely to move right or left given material variability because it takes no additional energy to move in either of those directions. The situation is similar to flipping a coin. Depending on how the coin is flipped, it will land heads or tails almost every time and yet there is a nonzero but very-small chance the coin will land on its edge.

The zero eigenvalue in the column system initiated a state transition, effectively eliminating the quasi-static equilibrium state and opening new pathways of motion that before the transition were impossible to follow. As eigenvalues decline toward zero, the orderly world of ideal linear systems breaks down. It now becomes much more challenging to accurately predict system behavior.

Removing the base of the column with an explosive charge would also induce a transition, but did an eigenvalue go to zero? No, not in the manner described above. The explosion suddenly creates a rigid-body mode that did not exist before that time. Rigid-body motion describes movements of objects that involve translation or rotation without deformation. *Rigid-body modes* appear in a buckled column made of brittle material where the transition caused the column to snap as it bends. At the moment it snaps the column is no longer attached to its base, and so it falls as a rigid body. We will return to this point below once we introduce a few more concepts. For now, we ask if anything can be done with this information to prevent a disastrous transition. Fortunately, there is.

The obvious thing to do during the initial stage of the nonlinear response in Figure 4.2, before the eigenvalues have changed significantly, is to reduce the applied load. If that happens while the system response is still mostly linear and elastic, the buckling transition can be avoided. We ignore microscopic material damage for now. Once system eigenvalues become zero, a transition to a new equilibrium state becomes inevitable, and, if the material properties are nonlinear, that transition is irreversible. If we can measure system eigenvalues rapidly with respect to the time it takes for the system to develop nonlinear behavior, there is an opportunity to prevent a transition from occurring. Note that all physically-realizable systems have the potential to respond nonlinearly over some input range. This includes electrical, biological, social, economic, and environmental systems. So serial monitoring of system eigenvalues can alert operators to impending transitions in these systems. Now let's examine the more-complicated example of Figure 4.3.

4.5   Equilibrium-state Transition: Ideal Arch

Consider the arching-plane structure shown in Figure 4.3. An old-fashion oilcan bottom is an example, where oscillating on-off pressure applied by your thumb to the bottom surface causes the bottom to snap inward and outward with an action that pumps oil out of the spout on top. This arch is purposely shallow, meaning its height is small compared to its span - the distance between supports that are hinged at the perimeter. The axisymmetric arching-plane can be represented by its cross-sectional profile across a diameter, as seen at the top of Figure 4.3. Lets apply eigen-analysis to this problem.

Applying a central downward force, the system initially responds with linearly-elastic compression of the material along the arch that also results in a vertical displacement v of the arch peak away from its resting

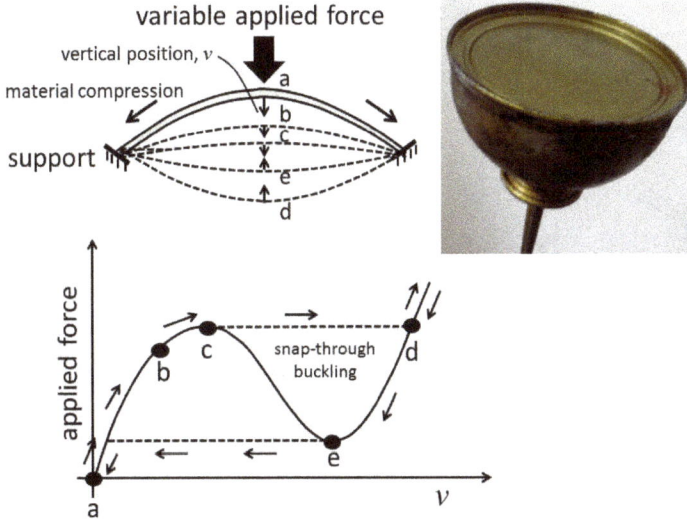

Fig. 4.3 Oilcan bottom (right) as an example of a shallow arch structure (left top) that undergoes reversible transitions to induce a pumping action. Initially, the arch moves vertically in proportion to the applied force. The phase plot at the bottom shows that by increasing the force beyond the linear range until the slope of the force-displacement curve goes to zero, indicating that an eigenvalue of the arch system has gone to zero, a snap-through buckling transition (horizontal dashed-lines) occurs. The solid curves are the incremental linear-system responses while the dashed lines describe the sudden movements during transitions. Notice that these reversible transitions follow different pathways. The displacement patterns shown are consistent with the hinged support that allows rotation.

equilibrium position at (a) in Figure 4.3. Linearity means that displacement is proportional to the applied force. Moving from (a) to (b), the system responds increasingly nonlinearly as the v-force curve in the phase plot flattens. Initially it requires substantial force to displace the arch downward. Once motion begins, less force is needed to generate the same incremental displacement until we reach point (c). Suddenly at (c), it takes virtually no additional force for the arch to invert and move to position (d). This event is called *snap-though* and it takes place dynamically [Pecknold *et al.* (1985)]. Let's freeze the action to see how eigen-analysis can help us explain this transitory behavior.

The arch has an infinite number of eigenfunctions (modes) and eigenvalues just like the column has; six are shown in Figure 4.4. Since the arch is symmetric about the central vertical axis, each eigenfunction can be classified as either symmetric or antisymmetric. When a central vertical

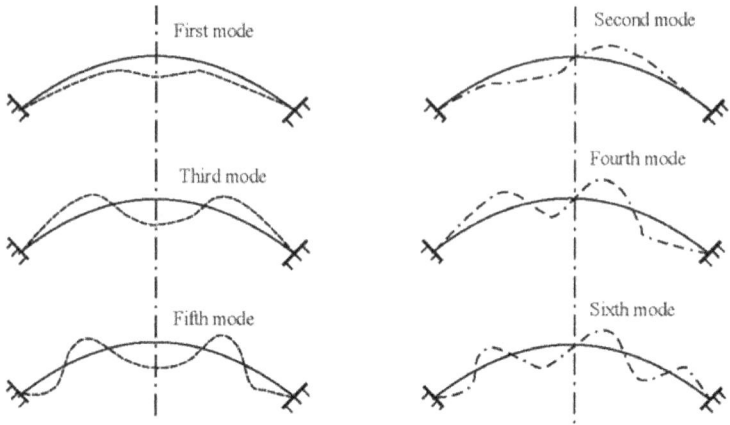

Fig. 4.4 Cross section of axisymmetric eigenfunctions corresponding to the first six modes describing downward motions of a metal oilcan bottom. The odd-numbered modes on the left are symmetric (about the central axis) eigenfunctions, and the even-numbered modes on the right are antisymmetric eigenfunctions.

force is applied to the central axis of a shallow arch, only symmetric eigenfunctions become activated throughout the operating range, including the snap-through event, and so the system deforms according to a linear combination of only symmetric eigenfunctions.

The mechanism that explains why only symmetric modes are involved with the geometric nonlinearity leading to a transition in a shallow arch is found by examining the mathematical model. There you will find that a snap-through transition is triggered before much stress builds within the shallow-arch shape. As the arch material is compressed before snap-through occurs, system properties change to lower the arch-system stability. We see the lowered stability as a reduction in the slope of the v-force curve between positions (a) and (c) in Figure 4.3 (eigenvalue is related to the slope) that eventually goes to zero. The changing eigenvalue is a warning of instability. The rate at which the eigenvalues for low-order symmetric modes approach zero is much higher than that of other modes, and so only a single-pathway for motion is available. Because just a few eigenstates are involved with this activity, we classify this situation as a low-dimensional problem in an infinite-dimensional system. Much more on system dimensionality and accessible modes will be given in Chapter 6.

We are now at point (c) in the experiment. An eigenvalue has gone to zero and the arch has begun the snap-through process. The slope becomes negative, meaning the arch is displaced further downward even if we reduce the force because of stored energy. Since it requires virtually no additional energy to move the arch further downward, the most likely occurrence is that the surface snaps from point (c) to (d) and the transition is complete. At (c) the geometry of the arch has reached a point where the energy stored as material compression is suddenly released. Significant additional force is required to move the surface downward beyond (d), even just a little, as now the arch material must be stretched.

If at point (d) the force from the thumb is reduced, the arch does NOT suddenly snap back to point (c). Instead it follows the solid-line pathway in the force-displacement curve of Figure 4.3 until it reaches the v-force curve minimum. At that point, the slope is zero as a symmetric-mode eigenvalue again goes to zero, and the reverse transition is initiated.

We specifically chose the shallow-arch geometry and an elastic material so that eigenvalues corresponding to low-order symmetric modes would go to zero first and the transitions would be reversible. Even if our thumb pressure is not exactly at the center of the oilcan bottom and we apply some force along antisymmetric modes, the shallow-arch geometry ensures that almost all of the system-property changes that occur during the nonlinear response phase of the motion affect mostly symmetric-mode eigenvalues. The antisymmetric-mode eigenvalues change very little. Therefore, the observed motion is symmetric. If material damage occurs from repeated snap-through motions but the damage slowly accumulates, the can may pump oil many thousands of times before the material cracks. We can declare the design as good for this application.

What happens if we select a steeper arch geometry so the ratio of height to span is increased as shown in Figure 4.5? The eigenfunction modes are similar to those in Figure 4.4 and, if we use the same elastic material, the effects of material properties on eigenvalues are unchanged. What changes is the nature of the nonlinear geometric response to force. The steep shape of the arch is more vertically oriented, so the applied force compresses the arch material more before the symmetric eigenvalue (slope of the v-force curve) goes to zero. The greater stress in the walls of the arch increases the rate at which eigenvalues for antisymmetric modes approach zero. In Figure 4.5, we can see an example where an eigenvalue for an antisymmetric mode goes to zero at point (b) and before point (e) where the symmetric-mode transition would have occurred. At (b), the movement is initially along the antisymmetric mode for the zero eigenvalue, which is either along the $+u$ axis (path through point c) or, along $-u$. Point (b) is the *bifurcation point* of this system. At this point, three solution paths become possible.

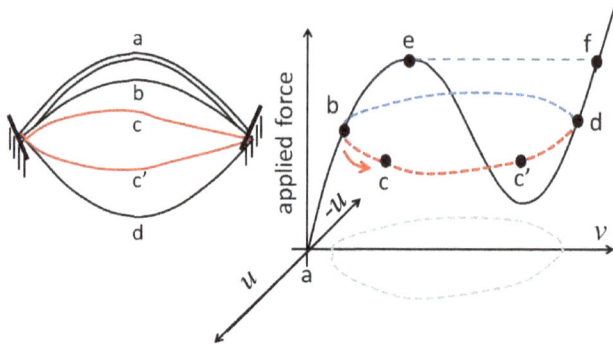

Fig. 4.5   Snap-though buckling in a steep-arching plane. With the steep geometry, the rate of decline of eigenvalues associated with antisymmetric modes is high. If eigenvalues for antisymmetric modes go to zero before those of the symmetric modes, as in this example, there will be a horizontal (u axis) component to the observed snap-through motion beginning at (b). Positive or negative lateral components are possible, as shown on the right, although in this case the positive path (through points c and c') was taken. Projecting the motion curves onto the u, v plane (dashed lines in the u, v plane), we find similar Ψ-shaped path options as shown in Fig. 4.2. The displacement patterns shown are consistent with the hinged support that allows rotation.

In addition to the $+u$ and u directions, the continuation of the original solution with symmetric displacements is possible in a perfect arch, but that path is highly unlikely.

In this specific case, we note that the motion takes place along the red curve. Immediately thereafter, all eigenvalues are nonzero and movement becomes a linear combination of all active modes that follow the path to point (d). The motion past point (d) is again the linear combination of all the symmetric eigenvectors. It is important to note that point (d) is also a bifurcation point in unloading the applied force. If we are unloading from a point above (d), when we reach (d), again the eigenvalue of one of the antisymmetric eigenvectors goes to zero and three solution paths become available. The steep-arch bifurcation-type buckling transition has similarities to the snap-through transition of the shallow arch: the transition to a new equilibrium state is reversible if the material is linearly-elastic. (The reverse path is not shown in Figure 4.5.) If the material properties of the arch are nonlinear, the transition will be irreversible. There are also important differences: The deep arch has three motion paths possible at point (b) in Figure 4.5 while the shallow arch has only one path at point (c) in Figure 4.3. In both cases, nonlinear system responses generate new equilibrium states once a transition is complete that may or may not be reversible.

At this point readers may be wondering why we are studying these mechanical systems in such detail. What could these systems possibly have in common with the economies and ecosystems described in Chapter 1? Mechanical systems are relatively simple and extensively studied. Most importantly, we are able to calculate their eigenstates mathematically. If they operate according to rules governing all systems, as we believe they must, then these examples have much to teach us about transitions in all systems, even economies and ecosystems where the eigenstates cannot now be computed.

Low-dimensional mechanical problems are just the tip of a large conceptual iceberg that offers new approaches for analyzing much larger high-dimensional systems; those too complex to model mathematically using methods known today.

4.6   Summary

Many simple low-dimensional systems can be modeled mathematically to reveal their eigenvalues. It is relatively straightforward to compute and interpret eigenvalues in terms of the material and geometric features we readily observe. Hence we can explain how system property changes are reflected in the eigenvalues. For example, a defect in an oilcan arch that changes its material properties with repeated use will be manifest in changing system eigenvalues that warn us of the looming failure. We conjecture that the predictive power of eigen-analysis applied to low-dimensional systems scales up to much larger systems because eigenstates are a fundamental descriptor of all systems.

We found that when the real part of at least one of the eigenvalues goes to zero, the system is undergoing a transition to a new equilibrium state with new system properties. The initial path of the response during a transition is always along the direction of the eigenvector corresponding to the zero eigenvalue. Once the transition is initiated, however, system eigenvalues continue to change. Zero eigenvalues become non-zero again, and the transition pathway involves system responses that are linear combinations of the eigenvectors as they exist at the current time.

4.7   Bifurcation Theory and Random Variability

Sudden transitions have been formally codified in the science and engineering literature as *bifurcation theory*. It provides us with an established mathematical foundation for explaining the nonlinear behavior under which columns and arches buckle. It also predicts transitions in other nonlinear systems like electrical and biological networks. Bifurcation theory may be applied to continuous or discrete systems to predict responses to input stimuli as eigenvalues approach zero. Our examples above are consistent with bifurcation theory but there are differences worth discussing.

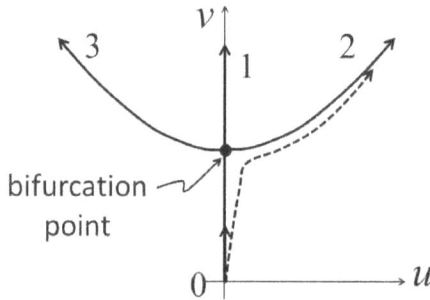

Fig. 4.6   Motion of the top end of the column in Fig. 4.2 that is subjected to an increasing load leading to buckling. The responses of a perfect column as predicted by bifurcation theory (solid lines) and that of a realistic column (dashed line) are plotted. The plot also describes motion in the (u, v) plane of the steep arch from Fig. 4.5.

To help explain, we modified the phase plots from Figures 4.2 and 4.5 and displayed them in the plot of Figure 4.6. Regarding the column example and loading its free end along the second eigenvector $(0, 1)$, the initial response is to move vertically in the direction of the applied force. This solution path (solid line in Figure 4.6) begins at the origin, point 0 in the (u, v) plane, and proceeds upward along the $+v$ axis. The column continues to move along this path as force increases until the response solution of the system reaches the bifurcation point, when an eigenvalue goes to zero and suddenly three solution paths become available, solid-line paths 1, 2, and 3 in Figure 4.6.

According to bifurcation theory, before a bifurcation point is reached, the top of the column is only able to move vertically. At the bifurcation point, however, the column can continue moving along path 1, the $+v$ axis, or a solution path that initially begins horizontally along either path 2, the $+u$ axis, or path 3, the $-u$ axis. If paths 2 or 3 are selected, along the first eigenvector $(1, 0)$, then a short time later the column will move as a linear combination of both eigenvectors. The paths in this 2-D phase space are upward and outward as the elastic-material column bends and falls. If you consider the magnitude of each eigenvalue as a measure of the energy needed to move along that eigenvector, then it makes sense that when the first eigenvalue goes to zero the zero-energy solutions open along $\pm u$. What ultimately determines the path taken?

To answer this, we must understand that the three pathways predicted by bifurcation theory only apply to the mythical ideal column that is perfect in every way. The path will remain exactly vertical only if the column material is absolutely uniform, its shape is perfectly straight and vertical, and the applied force is precisely central and oriented downward along the second eigenvector. The top of the ideal column will continue moving in the vertical direction past the bifurcation point. Solution paths 2 and 3 become available if we apply a very small lateral force in the direction of $\pm u$.

In the real world, the likelihood of there being a perfect column is extremely small, essentially zero. There are always enough imperfection and variability in a real system to make paths something like 2 or 3 much more likely, so path 1 beyond the bifurcation point is theoretically possible but extremely unlikely in real systems.

Realistic columns don't follow any of the three paths predicted by bifurcation theory; they follow untidy paths like the dashed line in Figure 4.6. The initial part of the solution path is not exactly vertical; there is always non-uniformity in the material and the load is not exactly vertical and may be slightly off the central axis of the column. All these imperfections cause the top of the column to move slightly to the right as it moves down under the applied load, in this case. Once the load is applied and the column begins to deform, it is pretty clear that if it buckles it will fall to the right.

Another important point about bifurcation theory is that it always predicts a point in the evolution of a system's equilibrium state where two or more paths are available. Consequently, bifurcation theory says nothing about cases like the snap-through transition of a shallow arch where one eigenvalue goes to zero but only one solution path is possible. Eigenanalysis covers all of these possible situations.

Variability in material and geometric properties (the eigenstates) especially at small spatial scales ensures physical columns never reach a bifurcation point. With bifurcation theory, designers can obtain general guidance about the nature of column stability that provides intuition about what might happen. Accurate predictions using bifurcation theory require a tremendous amount of detailed information to add to the mathematical models, information we almost never have for physical systems. It is important to understand the eigenstates of the column at all scales to accurately predict what will happen in a specific case. While one of the column eigenvalues does approach zero as the load increases and the stability of its equilibrium state is eroding, it is unlikely that that eigenvalue will actually equal zero, which is why the bifurcation point is never reached. Columns do fall, nevertheless, so how should we handle uncertainty when trying to predict events?

The dashed-line path in Figure 4.6 illustrates that real columns move along paths given by combinations of several or many accessible eigenfunctions. There is only one solution path that each column takes when responding to an applied force, from its initial equilibrium state to the final buckled state. Variability in material and geometric properties ensure the eigenvalues representing the nonlinear column system constantly change as the load increases. We don't know how to monitor the uncountable number of eigenstates in a real column. Local regions in the column material form subsystems with eigenvalues that suddenly approach zero when the loaded material fails in the form of cracking, crushing or yielding. This local transition may not induce a transition in the whole column but it can modify other eigenvalues of the column, reducing stability and increasing susceptibility to a future transition. When enough column subsystems fail, the column will fail. The future behavior of a whole system can be predicted by monitoring some of its subsystems, but which ones and at what scale?

A strict reductionist would say we need to measure every detail of that column and its environment to model the situation exactly in order to accurately predict the fate of the column. That course seems unwise for the very complicated engineering problem of predicting the behavior of nonlinear systems having an infinite number of eigenstates that can each change over time.

A strict non-reductionist acknowledges the unlikely possibility of knowing every detail of a column that might play a part in its structural stability, and so asks why bother with mathematical models at all since some amount of model-based prediction uncertainty is inevitable. We should just build and test many columns and use the scientific method to test updated hypotheses suggested to us by the data during the course of experimentation. This might be expensive to do but decisions made from these observations are always based on real-world experiences and unlikely to lead us astray.

The wisest problem solvers combine the strengths of both approaches. Experiments inform incomplete models that are verified and modified by additional experiments that inform and validate improved models, etc. Of course! It is hard to disagree with the scientific method. We further suggest that there is a conceptual tool that can facilitate both the reductionist and non-reductionist approaches. That tool is eigen-analysis, which describes the nature of system properties and their temporal stability at all scales. If a system is known and low dimensional, so we can compute its eigenvalues directly, then trained engineers and scientists can interpret the results to know what to do next.

For more complex systems we must discover ways to monitor their eigen-values. Even if we can follow just a few of the smallest eigenvalues most likely to approach zero, or those with similar values that could form sub-spaces where the basis functions become coupled (as with the suspension bridge example in Section 3.4), eigen-analysis is helpful for assessing system stability without full knowledge of the details of the system and the environmental stimuli. Of course, the more we learn about a system, the easier it is to measure its eigenvalues using computational methods. Combining bottom-up synthesis models championed by reductionists with a search for reliable global indicators as championed by non-reductionists makes a lot of sense.

Eigen-analysis is used today as a method for designing or explaining linear systems that are operating near their stable equilibrium states. In reality, eigen-analysis is a much more general concept and a powerful tool for exploring properties of all systems with inherent nonlinearity.

4.8   Transitions Within a System Hierarchy

Just as columns and arches are composed of subsystems like bricks and mortar, they may also be components of larger systems like buildings and bridges. The behavior of a simple isolated structure can be easy to char-acterize when the environmental stimuli are known and controllable, but as these components assemble into ever larger structures the complexity of their ensemble behavior increases significantly. The equilibrium states of simple structures help us understand their role as components of larger building systems. Relating the stability of a structure at one scale to other scales is addressed by eigen-analysis.

Each system component can have an infinite number of equilibrium states, but fortunately only a few of them are needed to analyze the problem. Building components may be designed to respond as linear-elastic systems, and yet we know they also respond nonlinearly in practice. When behaving nonlinearly, support subsystems undergo irreversible changes in their properties at variable rates and they can fail if a transition occurs. Subsystem transitions may not affect the stability of the whole-system equilibrium very much. However, local failures can propagate to larger scales to quickly degrade the stability of the whole building. The spatial hierarchy of transitions within systems of systems is worth a closer look.

Buildings are designed to function safely under a dead load (the weight of the building and its permanent parts), a live load (that resulting from occupancy), and environmental loads (wind, snow, earthquakes). Each can vary significantly during the projected life span of a building. Design parameters specify the material and geometric properties that determine system eigenvalues responsible for achieving the desired equilibrium state of a building. The building remains in equilibrium as long as small load changes generate small movements. A well-designed and built structure is one that maintains acceptable system properties over its lifespan despite the inevitable evolution of its equilibrium-state properties. The variances of projected environmental stimuli and the associated rates of equilibrium-state evolution (wear and tear) necessitate consideration of safety factors.

For example, let's assume the strength of materials limits the applied load in a building, and we expect a maximum load of X tons. If we can afford it, we might choose to use materials that support $2X$ tons to minimize any material damage for loads exceeding the X ton limit. We have a safety factor of 2. Use of stronger materials is one (rather expensive) way to diminish damage from load variance, clever geometric design is another, and both influence system eigenvalues.

Consider the case where a building is damaged by excessive loading. We saw previously that if column loading exceeds its buckling load, one or more eigenvalue can approach zero. If that column collapses, it will be crushed if made of reinforced concrete or bent inelastically if made of steel. The load that this buckled column carried is now transferred to the surrounding columns. If those columns can support the additional load without also buckling, then, although a portion has collapsed, the building remains in an acceptable equilibrium state.

If the transfer of the load from a buckled column causes other columns to buckle, even larger forces will be transferred to those that remain. If all load-carrying elements within a structure are the same, then excessive loads may cause a quick succession of buckling transitions leading to the collapse of the whole building. What is the eigenview of this situation?

We must first understand that each eigenvector of the building has a contribution from every structural component in a building. Each connected structure in a building experiences the forces applied to all other structures. However, the different eigenvectors weight those component contributions differently. It is not like one set of columns is associated with one eigenvector and another set with a different eigenvector; every system component contributes to some extent to every eigenvector. Some eigenvectors spread their weights more or less evenly over all structural components of a building, while others concentrate their weights locally, perhaps near a column, with negligible contributions from other parts of the building. For that reason, when an eigenvalue approaches zero, the effects of the ensuing transition can be local or global. If the associated eigenvector weights its component contributions almost exclusively near a column, the immediate effect of a transition will be local. If the eigenvector equally weights contributions over the whole building, then a transition can result in global collapse. The different spatial scales associated with each eigenvector in a system is important in design considerations. Equivalent cases can be made for economic, biological, organizational, and environmental systems as was made for buildings.

Once a focal transition has occurred, its effect on the rest of the system can propagate to larger spatial scales if the load added to the remaining support structure exceeds design capacity. In that case, one zero eigenvalue can induce others in succession. A well-designed building distributes supports in ways that inhibit quick successions of structural failures, giving operators time to intervene. Failure of one column in a building with a small safety factor has a potentially greater influence on overall stability than failures with a large safety factor. Using stronger materials increases the size of building eigenvalues, which is a wise investment when long-term strength of materials is the most vulnerable factor in equilibrium stability.

If the construction budget is severely limited, it might be financially advantageous to instead build in sensors that regularly monitor vulnerabilities that can be quickly repaired before structural damage occurs. Engineers are motivated to finding key indicators of system stability for just this purpose.

In Section 5.15 of the next chapter, we discuss cancer as a progressive disease of malignant transitions in cells that become dangerous to an organism only if cellular transitions are able to sequentially progress and influence larger scales in the body. Preventative medicine is a search for biomarkers that indicate changes in human-system eigenvalues. The concept is universal and therefore translate to many applications.

In each example of this chapter, a decision needs to be made about the scale for monitoring changes in eigenvalues in order to design appropriate early interventions. If we only monitor eigenvalues at the smallest structural scales of a large system, we might be sounding the alarm unnecessarily at great expense and concern if small-scale component failures can self-resolve or become isolated. Monitoring only at the largest system level may not leave time for effective interventions. So not only must we find ways to monitor system eigenvalues, we must do this at scales within the system hierarchy that identify components most critical to the stability of an equilibrium state. In Chapter 6 we further explore the idea that variability in the the spatial extent of eigenvectors increases stability of the equilibrium state.

4.9   Rigid-body Motion

In our description of support-column failure at the beginning of the chapter we noted that nonlinear material behavior during a transition can cause the column to crack and fall to the ground instead of reversibly bending as shown in Figure 4.2. Once a column is no longer attached to its base, zero eigenvalues are suddenly created that enable rigid-body motions. *Rigid-body modes* allow movement of mechanical systems without deformation. The general concept of rigid-body modes has broad implications that we now discuss.

We know from the discussion above that measurable energy is required to deform an object, and we can be sure that the deformation process is engaging system eigenvectors with nonzero eigenvalues. However, if that object is mechanically untethered and, say, lying on a table or floating in space, it takes virtually no energy (ignoring the energy needed to overcome inertia and air resistance) to move it as a rigid body. For that reason, rigid-body modes are system eigenvectors with initially-zero eigenvalues. Six modes are needed to describe all possible rigid-body motions in space but only one mode needs to appear to make the structure fall.

For example, removing a wedge of wood from the base of a tree trunk will activate at least one rotational rigid-body mode. Once the tree is completely unattached then all six modes appear. If we then shape the wood

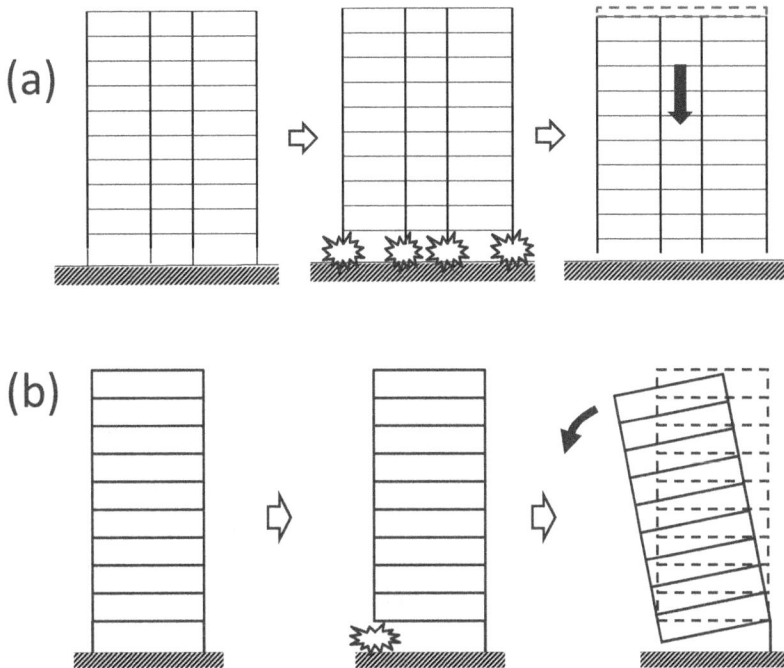

Fig. 4.7 (a) Diagram of building demolition by simultaneous explosive removal of all first-floor columns. Pre-detonation (left), explosions (center), and vertical rigid-body translation (right) causes the building to collapse in place. (b) Diagram of building demolition by explosive removal of the left-most columns activates a predominantly rotational rigid-body mode. This building falls to one side.

into a plank and firmly attach it to a wall in a building as a storage shelf, the zero eigenvalues all disappear because zero-energy rigid-body motions are no longer possible. But if the wood plank is unfastened from the wall, rigid-body modes reappear.

During intentional demolition of a building, explosives are used to create rigid-body modes. This process is fundamentally different from load-induced building failure. If the intention is to collapse the building onto itself, then the explosives are set to suddenly remove all the columns on the first floor simultaneously. As shown in Figure 4.7 (top), once the first-floor columns are eliminated, the upper stories suddenly acquire rigid-body modes, and the kinetic energy acquired by the ensuing gravitational fall acts to buckle the remaining supports on impact so the entire building effectively collapses in place into rubble.

The building in Figure 4.7 (bottom) is also transitioning; however, this first-floor explosion is designed to create a zero eigenvalue that primarily activates a rotational rigid-body mode with hopefully similar rubble-generating effects. These intentional building collapses can be looked at as transitions from a functioning building equilibrium state to a collapsed rubble state. In the case of a collapse under excessive loads, the transition is associated with some of the system non-zero eigenvalues approaching zero, but in the case of an intentional demolition, initially-zero eigenvalues are created as rigid-body modes. Are these different?

In both cases an eigenvalue is or becomes approximately zero. In the excessive load case, there might be a sequence of structural failures from localized support structures to larger scales. Similarly, for the explosive collapse of a building, as workers place explosive charges near support columns and set those charges for remote detonation, the eigenvalue of the building-workmen-explosives system is becoming smaller. We must consider the building along with the workmen and explosive materials to be part of the system if we are to see system eigenvalue going to zero. This should alert us to continually ask "what is the relevant system?" as we analyze problems so as to predict system behavior.

We conjecture that the causal relationships described by changes in eigenvalues that describe the mechanical transitions in a building generalize broadly to all systems. For example, consider forms of mass extinctions in population systems that have occurred.

4.10   Transitions in Population Systems

Since around 2006, the United States has been experiencing a phenomenon known as colony collapse disorder (CCD), a mysterious disappearance of honey bees from our natural environment. (For example, see [Johnson *et al.* (2009)]). It is always a serious matter when a species population suddenly goes into rapid decline, but honey bees are a species that play an essential supporting role in human agriculture, and therefore a CCD transition has a chance of propagating into large-scale human disasters. Colony collapse manifests as the rapid death of bees from a hive until there is no longer sufficient population size to sustain the colony system with its characteristic emergent properties, such as crop pollination. As a well-designed building can survive the loss of a few support beams, a stable bee colony is robust to the loss of individual bees. So the suspicion with CCD is that something out of the ordinary is reducing numbers quickly and with catastrophic consequences for several species.

The scientific literature provides evidence for new pathogens (bacteria and viruses), pesticides, and other environmental influences on bee health. How does one discover from many possible causes in this high-dimensional system[4] which effects are responsible for the CCD transition? Is it one source or a combination of sources? Society needs to solve this puzzle quickly, in time to halt the outward propagation of transitions before there are broad environmental and economic consequences. Responsible environmental stewardship is a complicated and time-sensitive business!

The big differences between species-stability and building-stability questions are the sizes of the systems affected, the investigational accessibility to system-component property monitoring, and the general approach to problem investigation. The latter point refers to buildings being designed by humans; therefore, we have the blueprints of the system to analyze, and we can understand, a priori, enough component properties to make reliable predictions. Buildings can be approached as a forward problem, meaning we can don our reductionist hats and conceptually assemble parts into models for analysis and virtual testing. In contrast, humans are not (yet) able to design a population of organisms. Each is a massively complex system of interacting complex systems, and our blueprints of nature's plans are still pretty sketchy. Living systems must be approached as an inverse problem, which means we must acquire many measurements on different individuals to establish cause-effect (stimulus-response) relationships. To manage all the unavoidable uncertainty within these systems, we apply statistical analysis to draw conclusions.

Research by Johnson et al. at the University of Illinois at Urbana-Champaign [Johnson *et al.* (2009)] invoked the inverse-problem approach to the analysis of CCD. They implemented their study by first selecting the appropriate scale for monitoring system activities that they argued was most informative. They surmised that the gut of each bee is the primary interface between individuals and their environment. So if CCD is the response of a nonlinear bee-colony system to a complicated mixture of environmental stimuli, they should be able to find evidence of such interactions within the cells of the gastrointestinal tract. They used microarray

---

[4]Simple systems like support columns and arches have infinite numbers of eigenstates just like bee colonies. A system may be considered "simple" when conditions dictate that it is functionally low dimensional; i.e., when only a few eigenstates out of many are actively generating the properties responsible for a behavior of interest. Bee colonies might be considered high-dimensional systems because a large number of eigenstates are thought to be responsible at all times for colony properties.

technology to look inside gut cells for genomic evidence, since nothing active happens within cells without leaving a genetic fingerprint (e.g., RNAs, proteins, etc.).

They examined bees from colonies affected and unaffected by CCD, and they found RNA fragments only in affected colonies that suggested a specific type of viral infection was involved. They did not find any evidence suggesting the cells were responding to pesticide toxins in their study population. It could be that a new virus came into contact with individuals that brought the virus back to the hive where it rapidly spread with high mortality. If generally true, then virus introduction may have ultimately caused one or more eigenvalues of the whole bee colony to suddenly approach zero. It is also possible that the viral infection coupled with other factors to trigger colony transition, and so the analysis continues.

Transitions within individual bees are rapid and too often the new equilibrium is a lifeless state. Transition of the colony seems to take more time as the virus spreads within the hive. If enough colonies are affected, the transition will propagate regionally and eventually globally to certainly influence human agriculture. At best, CCD is weakening the stability of human agricultural system equilibrium, making it vulnerable to future transitions initiated by a completely different trigger event. It can be difficult to define appropriate system boundaries when conducting such analyses.

Because we do not have a complete mapping of the "bee-colony system" as we have for buildings, we don't yet know how to measure bee-colony eigenvalues. However, tracking the residue of genomic activity is a promising indicator of changing cellular eigenvalues. It also seems that the gut cell is the appropriate scale for monitoring individual bees given their primary role (along with lung and skin cells) as interfaces to their environment. Genes control all cellular functions, and yet all changes in gene-expression that we might monitor do not indicate relevant eigenstate activities. The concern for entomologists and farmers is how to arrest CCD before it decimates the bee population and threatens our agricultural system. If we knew how to monitor changes in colony eigenvalues, we could predict CCD and perhaps intervene. Analogous to the collapse of a whole building from excessive loading of one floor, we believe CCD is an example of system eigenvalues approaching zero and initiating a devastating transition to bee populations that threatens to propagate to larger scales.

Is there a population-transition example analogous to a rigid-body-mode transition? We think there is. Consider that Earth, the only known living planet, has endured at least five *mass extinctions* during the billion-plus years that multicellular life has existed. The evidence for these events is found in the geological and fossil records. A mass extinction event is a widespread and rapid decrease in the diversity and abundance of planetary life. The biosphere undergoes a mass-extinction transition when the rate of species extinction far exceeds the rate of speciation.

The most famous and recent mass-extinction event occurred at the end of the Cretaceous period, the so-called K-T extinction of 66 million years ago. The geological record shows a 6.2-mile-wide meteorite collided with Earth, triggering a global ecosystem transition that rapidly eliminated about 75% of species larger than insects. The impact not only immediately eliminated many species, it also permanently changed environmental properties of the planet that supported macroscopic life; properties like the climate, sea levels, atmospheric gas composition, and average temperature. These long-term changes in the planetary ecosystem resulted in a new equilibrium state on Earth that offered advantages and disadvantages to the species that survived the event to quickly fill the available niches in the new ecosystem. For example, the K-T extinction may have eliminated all non-avian dinosaurs making it possible for mammalian species to expand their influence in the animal kingdom.

A meteorite collision is the equivalent of opening a rigid-body mode on the terrestrial ecosystem, as if the life-support rug was suddenly pulled from under all life forms simultaneously but with variable species-specific consequences. There seems to have been an unlikely sequence of events occurring in the solar system before the collision that, if it was closely followed today, might have been predicted in time to avoid a collision. In essence, an eigenvalue of the solar system was approaching zero, even if the Earth itself was unaffected until the strike, but the changes are only apparent once you consider system activities at the appropriate scale. What changes?

In 1950, Jan Oort, a professor at the University of Leiden in the Netherlands, suggested that comets originate from gravitational disturbances in a vast cloud of rocky ice balls that surrounds our solar system [Oort (1950)]. Our sun and the inner planets are at the bottom of a gravitational well. This well is surrounded by a ringed ridge of weakly tethered pieces of ice and rock. Once in a while one of those pieces is jostled in the direction of our sun causing it to fall down the gravitational well toward the sun. While we have never seen much of the vast junkyard of rubble in what is now called the Oort cloud outside of Neptune's orbit, its existence fits perfectly with centuries-long observations about how comets are introduced and periodically return. It may be that an icy comet falling toward the sun struck a rock in the asteroid belt, and a piece of that rock became the meteorite that struck Earth.

However it happened that a large rock came into Earth's orbit, the strike opened rigid-body-modes as the biosphere transitioned. Fortunately, the complex system of life tends to be highly adaptive to changing environmental conditions, so life eventually recovered in many new forms as dictated by the new environmental equilibrium state.

We can imagine parallel scenarios where eigen-analysis might help us track changes in the equilibrium states of other large complex systems technological, economic, governmental – to predict future transitions involving rigid-body modes in these systems. Our imaginings today may seem fanciful conjectures that are difficult to prove until we find reliable ways of monitoring system eigenvalues as readily as we can for simple mechanical systems.

4.11   Transitions in Systems of Granular Materials

In the examples of mechanical structures above, each system was described as a collection of firmly-connected components that form solid objects designed to minimize movement when subjected to environmental forces. Each structure expresses nonzero eigenvalues corresponding to its geometry and material properties that describe that systems ability to resist deformation. If a portion of the structure was to fracture and become separated from the whole, rigid-body modes with zero eigenvalues would appear for that portion, and the nonzero eigenvalues for the whole structure would become smaller as the ability of the system to resist movement is compromised.

Granular materials are a different type of structural system. They are collections of untethered particles that interact through gravitational and electromagnetic forces to generate characteristic emergent properties, and so they qualify as a complex system. Particles independently possess the properties of a solid that resist deformation (nonzero eigenvalues); however, those same particles are also free to move among neighbors because each possesses untethered rigid-body modes (zero eigenvalues). These simultaneous eigen-features give granular materials unusual properties.

For example, masses of sand grains on a beach will support our weight like other solid materials. However, when we pick up a handful of sand and let it slowly fall between our fingers, the grains flow like a fluid. Granular

Fig. 4.8   Examples of granular materials.

materials are neither solids nor fluids but share some properties of both. These features qualify granular materials for consideration as a fourth phase of matter, in addition to solids, fluids and gases.

Particles in granular materials must be large enough not to be primarily governed by thermodynamic forces. For example, molecular systems like air and water are not granular materials even though they too are complex systems composed of interacting particles with emergent properties. Familiar granular materials include beach sand, jelly beans, agricultural grain, and a ball pit in a childrens play area (Fig. 4.8). Each is fun to experience in part because of the sensation of the dual liquid-solid properties on our skin.

The interactions among the solid particles in a dry granular material are dominated by highly-dissipative inelastic collisions from frictional losses. Particles in contact can deform and even fracture according to their material properties. Interactions among neighboring particles follow simple mechanical laws but the emergent behavior can be very complex. We give a detailed modeling example in the next section.

A situation familiar to Midwest farmers is grain flow from a silo. Grain can be flowing out of an opening quickly at one instant and then suddenly stop. If you were to carefully inspect the grain you would see the particles have formed a solid arch – a jam – that inhibits flow. In essence, the granular particles have changed locally and suddenly from having properties of a liquid to those of a solid. Grain jams are an emergent property of highly-dissipative macro-particle flows through a narrow channel. They are more likely to occur in particles of variable sizes and shapes where inter-particle frictional losses are greater. Jams do not occur with pure inviscid liquids like water except through temperature changes; jamming is an emergent property of granular-material systems not shared by other phases of matter. A 2-D model simulation of different granular-material flows is illustrated in Figure 4.9, where it was shown through simulations how a solid particle arch could suddenly form [Alonso-Marroquin (2008)].

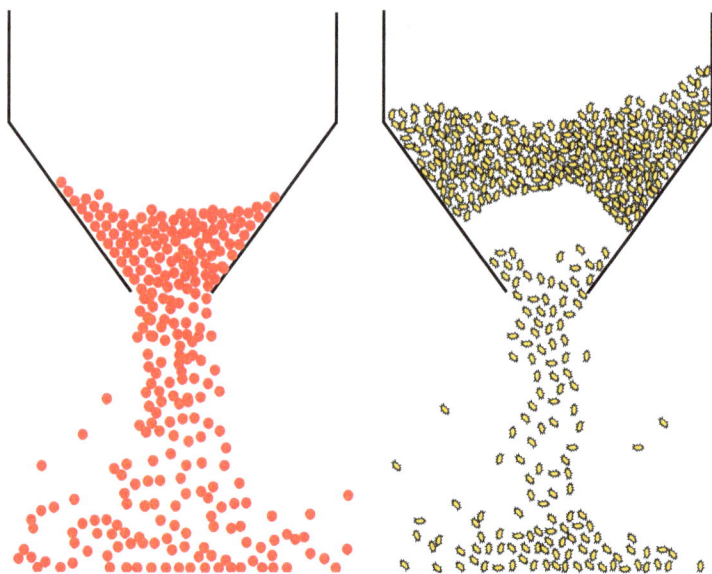

Fig. 4.9   Particle flows are simulated in 2-D models of granular materials. Eight seconds into the flow, the disk-shaped particles on the left are flowing smoothly, while the irregular polygons on the right form a solid arch jam. Drawing inspired by [Alonso-Marroquin (2008)].

The granular-material examples shown in Figure 4.8 are familiar to many of us. However, the general concept applies broadly to such unlikely systems as a stadium full of people exiting, luggage on a conveyor belt at an airport, or traffic on the highway. As in a grain silo, large numbers of densely-packed particles (people, luggage, vehicles) of variable sizes and shapes form self-organized systems with complex emergent properties despite there being a simple set of individual interactions. On the highway, these interactions generate highly dissipative flows when vehicles are confined to a narrow passageway that can suddenly form rigid internal structures that jam flows.

If you live in a city with a rush hour, you have experienced traffic jams and wondered in your frustration why high vehicle density at an exchange seems to bring traffic to a standstill for what seems to be no apparent reason. We say no reason because, when you reached the bottleneck, you seemed to sail through unimpeded. Hmmm...

In drivers' attempts to prevent collisions in high-density vehicle flows through a single-lane exit ramp, driver behavior effectively increases "particle friction". Individual drivers operating independently and with varying degrees of road vision and reaction times strive to avoid collisions by reducing their own vehicle speed to match what they perceive as a safe reaction time given the conditions. The concept of *autonomous vehicles* now being explored is an effort to reduce reaction variability within the self-regulated highway system to significantly reduce traffic jams. Relying on maps, sophisticated radar sensing, and a list of rules of interaction, robotic control of a "self-driven" vehicle can respond much more quickly and uniformly than human drivers. Uniformly-short response times allow for higher speeds and fewer jams while maintaining safe conditions.

The variability of component interactions among granular materials is not usually controllable. Yet these systems can be modeled using what we know to predict emergent properties including equilibrium-state transitions that are readily observable in controlled experiments. These self-organizing collections of interacting macro-particles are not well described by the established mathematical machinery of thermodynamics and statistical mechanics. Without reductionist tools, how can we characterize the behavior of nonlinear systems that result in unusual transitions like jamming? We might choose to develop a parallel mathematical analysis to statistical mechanics, as some are attempting now, or we can take this opportunity to explore non-reductionist methods. We have explored a few options with the help of a computational modeling method called Discrete Element Method (DEM). Regardless of the approach taken, the goal should be to search for indicators of eigenvalue activity.

4.12   Modeling Transitions in Granular Materials

We begin with a brief history of DEM in the context of its original application. This modeling technique was developed in 1970's to model the behavior of underground jointed rock. Underground rock masses are not continuous media as one might assume; they are blocks of rock separated by interfacing joints. Tectonic movements of the earth's crust over millions of years have fractured the solid rock in different directions resulting in the jointed rock masses that can be modeled as a granular material. The components of this system are the separate blocks that interact by moving along highly-dissipative jointed planes. Despite size differences, jointed rock and sand share similar emergent properties from the systems point of view.

Underground tunneling and mining activities require excavations through jointed rock. These excavations can disturb the equilibrium state of the block system that can lead to sudden transitions like roof collapses in mining tunnels. The earliest versions of DEM were developed specifically to model this behavior during the process of underground excavations. Recent versions of the algorithm were expanded to enable modeling of other types of granular materials [Ghaboussi and Barbosa (1990)], including those in Figure 4.8, which are the focus in the following modeling illustration.

Particle shapes and interactions are idealized in DEM models of granular materials. Nevertheless, the interactions they generate provide close approximations to the granular systems we observe. Particles are modeled in 3D as polyhedrons with flat surfaces and sharp edges and corners, although particles like grain often have smoother surfaces. In addition to shape, particle deformability influences these interactions. Each particle is assumed to be a rigid body in DEM connected to its neighbors through deformable springs located at contact points. As the springs deform, it appears in the model that adjacent particles penetrate each other (by a very small amount), which does not occur in reality, of course, without fracturing the particle. We remember what George Box said (See Section 4.3). DEM models do not try to precisely replicate every particle interaction, but they nevertheless accurately predict ensemble behavior, which is the modeling objective. In essence, an accurate model of a hypothetical system is created that behaves very much like real systems that we might observe. We are ready to simulate granular system dynamics, which are the time-varying movements of the granular particles.

The dynamics are modeled by a series of sequential snapshots of particle positions separated by tiny increments of time. The positions of each particle are known at the beginning of a time step. The initial task is to determine all contact areas on each particle. This is the most time consuming part of the simulation. Second, all forces acting on each particle are computed; these include external forces from gravity as well as those internal to the system from inter-particle and container surface contacts. Third, from these forces and the laws of mechanics, the positions of each particle are updated at the end of the time increment. The procedure is repeated for many time increments to generate a sequence of particle-position maps (movie frames) documenting changes in the granular system. This is an example of tangent or incremental property modeling that was illustrated in Figure 4.1.

Six time frames from a DEM simulation of granular-material dynamics are shown in Figure 4.10. We began with the static equilibrium state of the particles in a transparent box that existed just prior to panel (a). At the time of panel (a), two walls of the container have suddenly been removed, initiating a system transition where particles flow toward the viewer until they are again at rest in a new static equilibrium state. The system is in static equilibrium both prior to panel (a) and after panel (f). In equilibrium, a small perturbation to the particles produces a small response, like reaching in with your hand and gently moving the top particles. The equilibrium states are static when all granular-particle motion is insignificant.

The value of this relatively simple simulation is that it enables us to describe the eigenvalues of the system at every scale and at all times before, during and after the transition without having to compute the system eigenvalues. Let's look closely at the eigenvalues at various spatial scales as a function of time during the transition.

4.13 Interpreting Models of Granular Materials

All particles in contact with the walls just prior to panel (a) in Figure 4.10, at the instant before the walls are removed, are stationary because all forces acting on each particle, including gravity and neighboring-particle and wall contacts, balance each other. When the sum of all forces acting on a particle (the net force) is zero, that particle is in static equilibrium. Of course, the net macroscopic force is zero for every particle in the system prior to panel (a), so the whole system is in static equilibrium. There are nonzero eigenvalues that describe the geometric and material properties of this system at every scale from the total mass of particles to each individual

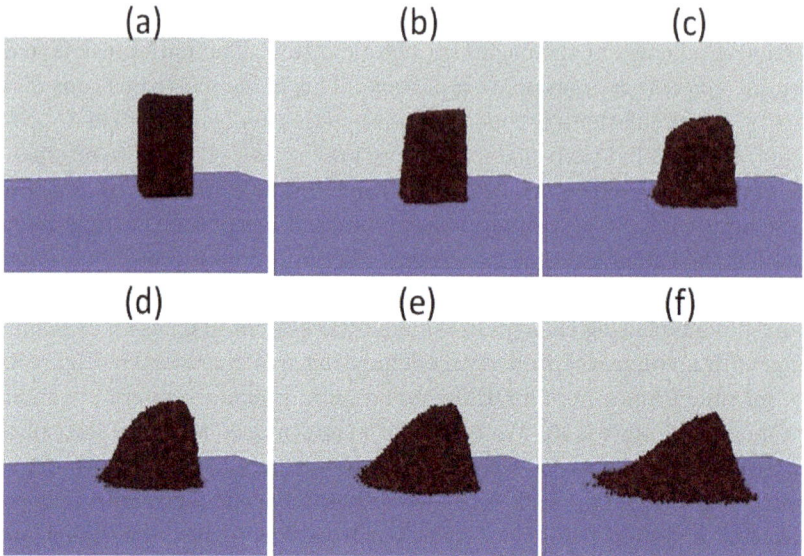

Fig. 4.10   Panels (a) through (f) illustrate the progressive stages in a computer simulation of granular material flow using a discrete-element modeling method. The granular materials are polyhedrons initially at rest in a glass box bounded by four transparent vertical walls. At the instant of panel (a), the two vertical walls for the surfaces you see are suddenly removed while the two back walls remain in place. Wall removal initiates a transition where granular particles flow toward us progressing in time from panel (a) to (f) to show the changes in system state during the transition. Just prior to panel (a) and just after panel (f) the system is in two different static equilibrium states. [Zhao *et al.* (2006); Zhao (2008)].

particle in the mass. We can compute all of these eigenvalues if we so choose because the DEM computer model contains every relevant detail of the system, and yet the system is simple enough to intuit the changes we are about to describe.

The instant that the two glass barriers are removed, barrier-contact forces on each of the outermost particles are suddenly set to zero while other forces acting on those particles are unchanged (Fig. 4.11). Consequently, the outer layer of particles in contact with the barrier acquires a net force, and those particles begin to move outward. Importantly, all of the other system particles in panel (a) remain motionless in static equilibrium. Let's freeze the system at this critical time to look at the eigenvalues characterizing different scales.

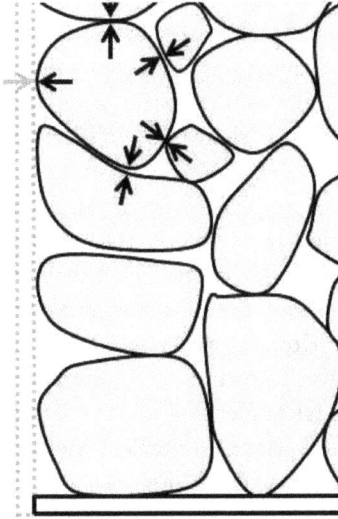

Fig. 4.11 Particles are shown from the lower-left corner of panel (a) in Fig. 4.10 near the left wall that was just removed (shown above in gray). Arrows show the forces on one particle, which were balanced at equilibrium, and are now unbalanced as the wall is removed.

Some eigenvalues for the whole granular system have gone to zero because a structure maintaining the static equilibrium state was removed. The eigenvectors of those zero eigenvalues are oriented outward in a direction perpendicular to the missing walls. At that instant, the outermost particles each suddenly acquire rigid-body modes according to their position in the system geometry and a combination of particle-contact and gravitational forces at the time of wall removal, and their initial motion is in the direction of the zero system eigenvectors. Had a particle experienced any linear-elastic deformations before the transition, they will spring back to their original shape during the transition, and so the individual particle eigenvalues are unchanged. In contrast, the eigenvalues of the granular system do change with the particle motion. What happens next?

As the outer particles move, the net force on the next layer of particles becomes nonzero and oriented outward, and so they too begin moving as each particle acquires a rigid-body mode that didn't exist just a moment earlier. Particle motion changes the properties of the granular system; consequently, the eigenvectors and eigenvalues of the granular system (not the individual particles) are also changing. Although some system eigenvalues remain zero throughout the transition period, the eigenstates are all in flux through the transition. We could sense which eigenvalues are zero at any instant by tracking the moving particles that must follow a path given by a linear combination of eigenvectors having zero eigenvalues. These falling-particle paths are zero-energy, rigid-body mode trajectories – the paths of least resistance.

The process continues with the next layer of particles, etc., until the granular system transition is complete, the net force on each particle is again zero, and all particles are at rest in a new static equilibrium state. After panel (f), properties of the granular system in its new equilibrium state are reestablished. How do we know? Once again, a small stimulus generates a small response according to the new properties.

Interestingly, we can tell something about frictional losses (due to particle shape, size variability, surface properties) occurring as particles interact during the time they are moving. In Figure 4.10, particle friction was relatively high because the initial granular system shape is still recognizable after the transition is complete. If frictional losses among particles were very low, as we might find for oiled ball-bearing particles of the same size, the falling particles would continue to move outward until they were just a monolayer across the horizontal surface. We gained significant information about this process by interpreting observable changes in terms of eigenvalues as the transition unfolded.

Having this detailed model of the granular system at all times between the two equilibrium states would have enabled us to mathematically compute all of the eigenstates exactly if that was necessary, but that is a long tedious computational process. We could compute eigenvalues at every scale, from the whole mass of particles to the particle layers and down to individual particles. All we need is each and every detail of every particle in the system at all times. If we have all this information for a physical system with negligible errors, and that is asking quite a lot, we would say the system is exactly known. And yet, even without all of this detailed information we are able to make two general observations about this and any system near its point of transition.

1. When at least one system eigenvalue goes to zero, the system transitions from one equilibrium state to another. When a transition is possible, small perturbations can cause large changes in the system.
2. At the beginning of the transition, the initial system response is always along the eigenvector corresponding to the zero eigenvalue,

or a linear combination of eigenvectors corresponding to the zero eigenvalues.

We believe these two observations apply to all complex systems.

Observations gleaned from this simulation can be extended through analogy to natural granular-materials phenomena, like avalanches occurring in snowy mountains. Snowflakes are sticky granular materials that accumulate on a slope. As they settle on the ground, their natural initial state is to form a jam, a granular-material solid. If allowed to pile up, then when some of the eigenvalues of the snow mass approach zero, that system undergoes a transition to a new equilibrium state. At the onset of the transition, the granular system effectively changes from its initial solid phase to a liquid phase that flow (avalanche) once any small stimulus, even a skier going over the slope, comes along. In the mountains of ski resorts, the authorities fire cannons so the resulting acoustic vibrations will trigger any incipient avalanches before skiers do.

Fig. 4.12 Avalanche and mudslide as examples of granular materials undergoing a sudden transition. Both photos are from Shutterstock.

Landslides are a similar type of sudden transition in soil systems. Soils are sticky granular materials of variable size and shape including clay, silt, sand, gravel and rock. Landslides can occur when rain soaks through the ground and increases the unit weight of the saturated soil. At the point when a landslide or slope failure is initiated, at least one of the system eigenvalues is zero because motion occurs without applying additional energy. At least one eigenvalue remains zero during the ground movement until the slide stops. The new position of the ground mass after the landslide or ground failure is the new equilibrium state.

4.14   Earthquakes as System Transitions

Our knowledge of earthquakes has increased significantly since the introduction of advanced oceanographic measurement technologies in the 1960s.
We know that most earthquakes occur at plate boundaries in the Earth's
crust (see Fig. 4.13). The crust is the outermost rock layer of our planet
that lays over a hot (500 - 900°C) viscous fluidic mantle covering the core.
The crust is broken into plates that move very slowly, at about the rate at
which fingernails grow. *Tectonics* refers to the dynamic processes within
our planet that direct the movement of those plates. The components
of this system are the floating crustal-plates that move to interact with
each other and the mantle at their boundaries. The system is driven by
a still-unknown combination of input sources, including heat convection,
material-density variations, and Earth-Moon tidal forces. The responses of
the Earth system to these source are plate movements yielding phenomena
like volcanos, earthquakes, and mountain building.

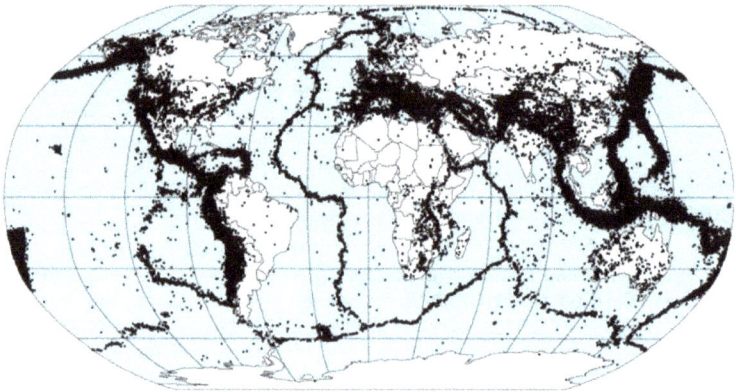

Fig. 4.13   The map of earthquake epicenters from 1963-1998 (358,214 events) mostly
coincide with the crustal plate boundaries (Source: USGS via [NASA (2004)]).

Crustal plates move usually just centimeters per year over a planetary
circumference of 24,901 miles. Yet the effects of this small motion over

millennia have completely reshaped Earth's surface topology. Interactions at plate boundaries can be classified into four categories: (a) divergent boundary interactions occur when plates move away from each other and (b) transform boundary interactions occur when plates move horizontally parallel to each other. Convergent boundary interactions are of two types: (c) collisional where the plates are moving towards each other at the same level and (d) subductional where the plates are moving toward each other but one has slipped below the other.

The mid-Atlantic ridge is probably the best example of a divergent plate boundary. A ridge of mountains on the ocean floor lies over this boundary between the North American and Eurasian plates in the North Atlantic and at the boundary between South-American and African plates in the South Atlantic. The ridge runs almost the full length of the Atlantic Ocean in the north-south direction (Fig. 4.13). It is thought that convective forces in the hot mantle below are pushing up the ocean floor crust causing these plates to spread. As they spread, magma from the mantle rises to reach the sea floor as lava, adding new material to the ridge. A rift valley running along the full length of the mid-Atlantic ridge is the actual plate boundary.

A magnificent example of the collision-type convergent boundary is that occurring under the Himalayan mountain range where the Indian plate collides with the Eurasian plate. The Himalayan range runs west-northwest to east-southeast in an arc 1,500 miles long with peaks as high as 29,000 ft in elevation. As these plates collide, the collapsing crustal plate is squeezed upward to form a mountain range that is clear testament to the amazing power of collisional plate interactions.

Most earthquakes occur at the transform and subduction boundaries. We will concentrate on earthquakes resulting from plates moving parallel to each other (transform boundary). Most of the earthquakes in California along the San Andreas Fault, at the boundary between the Pacific and the North American plates, belong to this category. The Pacific plate is the worlds largest, occupying most of the Pacific Ocean. About 90% of earthquakes and most volcanos occur on the edges of the Pacific plate called the "Ring of Fire." The Pacific plate is moving in the northwest direction relative to the North American plate as it rotates around a point south of Australia. The movement of the plate is just 2 to 3 inches per year, but this motion results in all four types of plate boundary interactions along its edge. These interactions have resulted in many earthquakes and volcanos during just the last 50 years as indicated in Figure 4.13.

Along the San Andreas Fault in California, a transform-type of strike-slip interaction is occurring (Fig. 4.14). This plate movement causes shearing strain at the plate boundaries. The plates cannot move freely with respect to each other because they are locked at the boundaries by frictional forces. When these shear strains reach a certain value (normally called the shear strength) suddenly the plates move with respect to each other causing an energy release that results in an earthquake. Of course, this type of sudden energy release does not occur along the full length of the fault all at once.

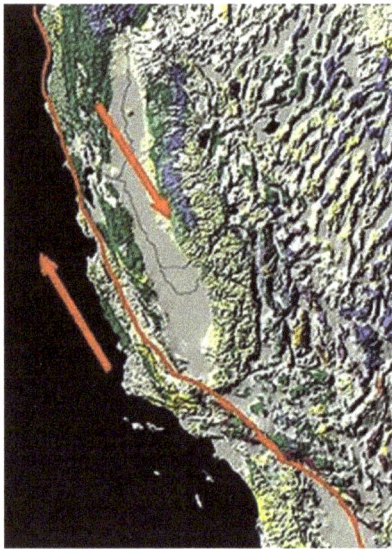

Fig. 4.14 A satellite view of the San Andreas Fault in California (red line), where arrows indicate the relative directions of the two plates.

The plate boundary is not a straight line and the fault is not a single boundary. There are many irregularities and multiple smaller faults. A rupture may occur at a boundary point along a small portion of the fault, resulting in a minor earthquake. When the fault ruptures, it releases shear strains that are transferred to the other parts of the fault, just like a failing support column distributes its load to neighboring support structures in a building. Sometimes the transferred shear strain is beyond the shear strength of its neighbors, which causes them to also rupture and a large earthquake is generated. In rare occasions fault ruptures may trigger a succession of ruptures leading to a very large earthquake along almost the full length of the fault, as in the 1906 San Francisco Earthquake. Strain release along the San Andreas Fault in 1906 caused a large relative displacement that could be observed at the roads, walls, and fences that crossed the fault. At Tomales Bay the offset along the fault was about 21 feet. Geologists believe that the total accumulated offset along the San Andreas Fault from the past earthquakes over the 15 - 20 million years since it came into existence is about 350 miles. They have determined this figure by matching the geological features on the two sides along the fault.

To explore transform-boundary interactions generating earthquakes in more detail, we now ask readers to conduct an experiment using your two hands. The principles of the experiment you are asked to perform are illustrated in Figure 4.15. They apply equally well to sliding hands as they do to sliding crustal plates.

4.15   A Simple Experiment

The eigenview of plate tectonics tells us an earthquake is a sudden equilibrium-state transition occurring when some system eigenvalues go to zero. To see this, we ask you to conduct an "exercise" where two meanings of the word both apply. Firmly press your dry palms together in front of your chest while trying just a little to move your left hand up and your right hand down. That is, apply a large compressive force and a small shear force. The shear stress you are generating (small at this time, hopefully) is the up-down force divided by the area of your palms. Your palms represent the boundaries between two plates and the friction between your palm surfaces is analogous to that at the crustal-plate fault. If you are successful at this couch experiment, your palms have not yet slid as in Figure 4.15(a).

Now apply a bit more shear force while maintaining the strong compressive force. You will notice the skin of your palms distorting a little, and, if they still do not slip, you will have successfully generated a shear stress less than the shear strength of the boundary. This state looks somewhat like panel (b) of Figure 4.15. Shear stress, roughly speaking, is the amount of lateral force on your skin divided by the area of your palms in contact. By now your arms are getting tired and beginning to shake with exertion, but your quest for knowledge is strong so you carry on. With your last remaining strength, increase the up-down shear force until your hands slip a little bit, panel (c), as the applied shear force exceeds the shear strength. Increasing the shear stress further, your hands will completely slip.

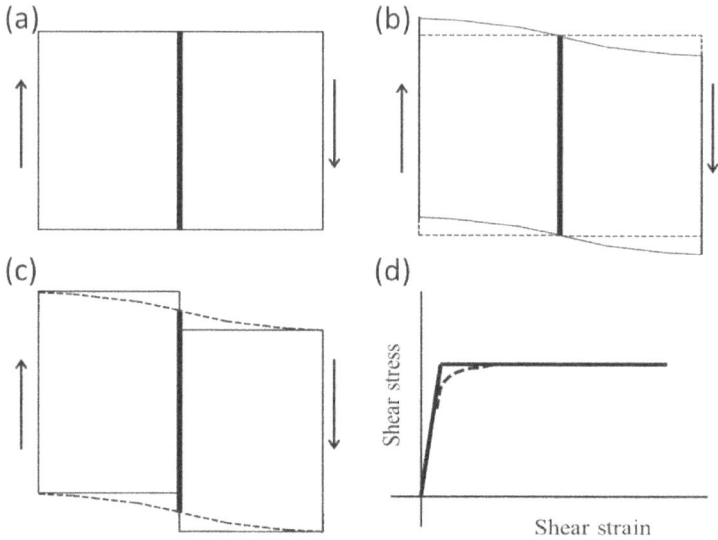

Fig. 4.15   A diagram of an idealized strike-slip fault at a crustal boundary.

Catching your breath, you realize that before your hands slipped your hand-arm combination was operating as a linear-elastic mechanical system. The shear stress is proportional to the shear strain, and the constant of proportionality (the slope of the strain-stress curve) is determined by material properties of your skin: its stretchiness and surface roughness. A plot of the stress-strain curve traces a straight-line path in the linear range from the origin of Figure 4.15(d) upwards with a steep slope. The steep slope tells us that little strain is produced even at high stress. In this linear range, if you slowly release the shear force while maintain the compressive force, the shear strain would follow the same straight-line curve back to the origin.

Ideally, your efforts would successfully maintain the linear-elastic behavior until your hands completely slipped. Since a force is applied along the vertical axis, once the eigenvalue of the system goes to exactly zero and your hands slip completely, the corresponding eigenvector must be along the direction of the applied shear force. We know this because the zero-energy path is along the eigenvector of the zero eigenvalue. This is seen in panel (d) of Figure 4.15 as the sudden solid-line break to the right in the stress-strain curve, showing the nonlinear response that takes virtually no additional stress to apply a very large strain. You also were applying a horizontal compressive force, but the eigenvalue along that eigenvector did not go to zero so ideally there is no horizontal motion (not shown in (d)).

In reality, your hands first began slipping just a little. That small loss of frictional contact led to more slippage until your palms were no longer in contact. So your hands followed the dotted line path in Figure 4.15(d). No eigenvalues for the whole system went exactly to zero as your hands began slipping, but the eigenvalues were getting smaller. This told you that stability of the equilibrium state was quickly eroding. There was, however, a small region of your hand surface where the eigenvalue went to zero and that small region transitioned (slipped). That slippage redistributed the stress from the region of failed contact locally, which changed the system slightly. The new system had another region where an eigenvalue went to zero, etc. We've seen this pattern before in the example of building supports, where a transition at one scale led to others at larger scales.

This is the analogous situation for earthquakes. Like in the hand experiment, the weakest part of the system – that most likely to cause a transition – is the frictional resistance to shear strain occurring at the boundary. However, failure is not all or nothing, but follows a spatial hierarchy. How high up the hierarchy a fault rupture propagated determines the magnitude of the earthquake. Often eigenvalues of Earth's *local eigenvectors* approach zero, which causes a *local transition* to occur, and yet overall the Earth remains in a stable equilibrium state. See the discussion in Chapter 6.

4.16 Summary

The new ideas introduced in this chapter and illustrated by mostly mechanical examples are related to highly nonlinear behavior associated with equilibrium-state transitions. These illustrations are essential for exploring larger systems for which we know many fewer details of their internal structures. We looked closely at what it means for an eigenvalue to go to zero and induce a transition. In the case of a column, we understand there are infinite numbers of eigenstates, but when we focused on the two eigenvectors for just one point on the 2-D column we still learned quite a bit. The column behaved as a nearly linear-elastic system for some range on input loads, which we defined as the designed operating range. Outside the operating range, we found nonlinear responses that could result in failures described as transitions. We idealized the column system to predict bifurcation points. However, real systems are more predictable if we are able to monitor eigenvalues at the appropriate scale in the system hierarchy. Monitoring systems throughout their functional lives, we can observe stages of equilibrium instability leading to failure as it occurs at different scales. This idea is just as true in medicine and the environment as it is in buildings. Selecting the appropriate scale for monitoring can alert us to changes with enough time to intervene. Without calculations, we learned quite a bit about column and arch dynamics that was expanded to analyze granular materials and plate tectonics. We believe these examples illustrate general principles true for all systems at all scales.

The arch examples showed another way to categorize infinite numbers of eigenstates in a way that predicts overall system behavior through interpretations of stimulus-responses data – the non-reductionist approach. The distinctions between system nonlinearity at various scales, and material and geometric nonlinearities, are important for predicting when transitions occur, for knowing if a transition is reversible, and, perhaps someday, for preventing transitions to undesirable states. We noted the importance of carefully considering what constitutes the system for the purpose of analysis and when interpreting eigenvalues. For example, the eigenvalues for the local eigenvectors along a small section of a fault line can go to zero, which changes the equilibrium state around the rupture significantly while negligibly influencing the equilibrium state of the Earth. That is, the spatial scale of influence of each eigenvector is critically important for prediction.

Transitions at small scales in large systems can propagate to have large implications for the system. Because it is difficult to draw boundaries around a system, it seems that all systems in the universe are connected. The question is how to use eigen-analysis to reliably predict the stability of the equilibrium states of these systems at every scale?

We discussed rigid-body modes where initially-zero eigenvalues appear to provide pathways to movements that require no additional energy. Rigid-body modes are important consequences of transitioning systems. They occur as a result of a larger system undergoing an erosion of its stability until a transition is triggered. Some systems, like mechanical structures, have a system-level eigenvalue approach zero but never equal zero. Yet a transition has occurred. Other systems, like granular materials, have eigenvalues at all levels going to exactly zero. Elements of the system can remain linear while others deform nonlinearly and undergo rigid-body motions. The concepts from granular materials are analogous to a broad and very diverse range of systems. The essential point to remember is that systems have universal principles that apply even to those too large and complex to be analyzed in the reductionist detail that we easily apply to simple mechanical systems.

Chapter 5

Biological Systems

5.1 Introduction

This chapter begins with analogous stories about two complex biological systems that are each composed of a network of interconnected components joining inputs to outputs. In both stories, an input stimulus propagates in time through several potential network "pathways" to generate an output response. The different pathways achieve similar responses but with variable efficiencies and costs. Some network functions are guided and others arise spontaneously without guidance. Networked systems in biology have evolved to provide countless numbers of cellular functions offering organisms the advantages of redundancy and multifunctionality as they struggle to survive in variable environments.

In the first story, Bob is a sixty-year-old baby boomer standing in line at a Rolling Stones concert in Chicago to buy a T-shirt. The shirt is being offered by a vendor at a table surrounded by many other concert goers also hoping to buy a shirt. Bob wants the T-shirt with the iconic picture from the cover of the Stone's Forty Licks album, although he is unsure where he might wear the darn thing given his current suit-and-tie lifestyle. He waits pensively during a set break at the outer edge of a cluster of people surrounding the table, thinking he could miss the next set of music if he is too patient. So he formulates a fair plan that minimizes his waiting time.

The engineer in Bob recognizes that this shirt-buying endeavor involves predicting the behavior of a complex system, where the inputs are money in hand and body location proximal to a vendor. The desired output is shirt acquisition in time to make it back to his seat for the next set. The components of the network are the steps that Bob takes through the crowd

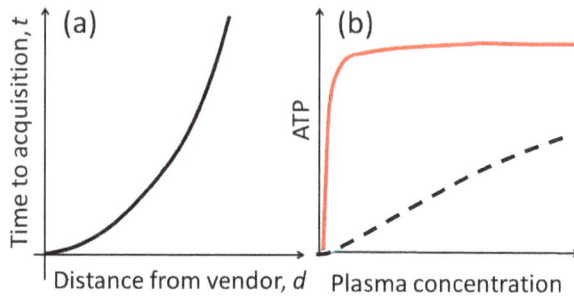

Fig. 5.1   (a) Bob's strategy for navigating the crowd that minimizes acquisition time is based on this simple input-output model describing how acquisition time varies with distance if one takes the first crowd opening to move. (b) The production of cellular ATP as a function of sugar concentration in blood plasma is shown for two cell types. Highly efficient energy-generating cells are represented by the upper curve, and lower-efficiency cells by the dashed-line curve.

toward the vendor table while being respectful of the personal space of others in the crowd. Knowledge of system properties can efficiently direct his movements, so Bob assesses the situation and poses the *stimulus-response model* given by the nonlinear curve of Figure 5.1(a). Distance d and time t are the state variables of his phase-space model. The curve predicts that if you are far from the table, as Bob is currently, it will take a disproportionately long amount of time to get to the vendor if the plan is wandering around taking any opening in the crowd to move. Bob decides his best strategy is to first press directly toward the vendor table. Then when he is close, where the crowd is most dense, he will maneuver around the table to find an available clerk. Of course, things can always go wrong because of the randomness of the crowd, but Bob feels that on average this strategy will help him achieve his goal of minimizing the acquisition time. His path is shown in Figure 5.2(a).

In the second story, a cell in Bob's body is attempting to gather molecular resources from the blood supply to fuel Bob's movements through the crowd. Functions within each cell are generated by coordinating processes occurring in several parallel cell-system networks, but we will focus on one specific component of the metabolic network. The inputs are nutrients (carbohydrates and oxygen) and the output is chemical energy stored as adenosine triphosphate (ATP) molecules (Fig. 5.1(b)). ATP production involves a sequence of chemical transformations. Complex carbohydrates broken down (catabolized) outside the cell into small sugar molecules (monosaccharides) are easily passed from capillary blood plasma through the interstitial

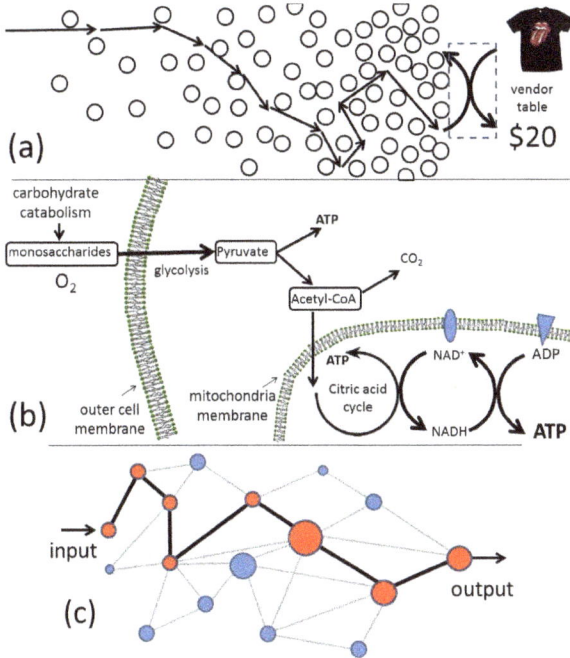

Fig. 5.2 (a) Illustration of Bob's piecewise linear path through the crowd as guided by the stimulus-response model of Fig. 5.1(a). (b) A coarse-scale illustration of cellular ATP production within mitochondria from the catalysis of carbohydrates and oxygen. The stimulus-response relationship for this system is given in Fig. 5.1(b). (c) A general graph representing a coarse-grained network connecting one input node to one output node. It is coarse-grained because each node might represent an extensive sub-network. Nodes connected by dark-line edges indicate a currently active pathway. The sequence progresses in time from left to right. Lighter, gray-colored edges indicate less active pathways. Node size may indicate content size or influence on the process.

fluid space and outer cell membrane and into the cell as shown in Figure 5.2(b). Once inside, numerous chemical reactions transpire before the cell achieves its desired output – ATP production. ATP efficiently provides the chemical energy necessary to power cellular activities; it is not the only energy source in a cell but it is an important one.

Each chemical species formed in the cell and each position along Bob's path can be modeled as "nodes" in a network, as illustrated in Figure 5.2(c). The reactions transforming one chemical molecule into another or the action of Bob choosing a direction and initiating a step are each represented by lines connecting the nodes. The connectivity of a network can be visualized from the topology of the graph illustrating connections among system components. A graph is a tool for which users must assign physical meaning to the components. The lines are generally called network "edges," but specifically for metabolic networks they represent reaction fluxes. In both stories, a great many detailed events take place within each node, sometimes hundreds or thousands, yet we group them coarsely at our discretion into just a few nodes to summarize the most important steps in the sequence so we can address the specific questions at hand. *Graphs* are very general tools often applied to help visualize and eventually predict the behavior of networked systems.

Each step along the chemical sequence brings the cell a little closer to its objective of energy production, and yet each event also includes the potential for redirecting subsequent events along a different network path depending on the environmental signals a cell receives. Cells receiving signals from their environment and performing programmed functions can be directed during the process to take a variety of paths or perhaps several simultaneously. Hence cells "listen" to their environment when making "decisions" that determine their behavior. Although scientists anthropomorphize their explanations of cellular processes to be more easily understood by everyone, using terms like listening and decision making, these are actual mechanistic operations albeit devoid of cognitive processes.

Similar to cellular decision making, each step that Bob takes that brings him closer to the vendor table has the potential to modify his tactics when planning the next step. Depending on how people in front of him are moving, he may weave and spin to take an opening that gets him closer to the table. Bob's movement decisions, however, do involve cognitive processes informed by sensory data.

In both examples, having access to many paths through the network introduce opportunities and challenges for achieving goals in rapidly changing environments. An advantage of a network structure is to provide alternative solutions for achievement when one path becomes blocked. Our protagonists, Bob and the cell, need alternatives for decision making to be productive. Cells can employ several metabolic pathways simultaneously, while Bob considers several courses of movement after each step. It is the ensemble effects of these networked sequences that spawn emergent system properties enabling both biological systems to adapt to variable environments. Nevertheless, some cells will be more efficient at one form of energy production than others, and not all shirt-buying Rolling Stone fans will get back to their seats before the intermission ends.

The cell's energy production pathway and Bob's shirt-seeking path are both self-regulated processes. In the case of the cell, after sensing the available resources, DNA in the nucleus directs construction of the molecular machinery needed to carry out ATP production. Energy-producing molecular machinery is built once signals indicate that appropriate raw materials are available. A different metabolic pathway is activated in the cell depending on whether Bob eats a high-protein, high-fat, or high-carbohydrate diet before the concert. Normally functioning cells make decisions that allow the cell to quickly adapt and exploit the available resources.

Bob's path through the crowd is also self-regulated. His nervous system, built from DNA instructions and developed during an accumulation of life experiences of navigating crowds, generates instinctual movements that are modified by Bob's perception of what is happening in the crowd around him. His moment-by-moment pathway decisions ultimately determine how successful he will be at achieving his goal.

In both systems there is no central control from outside the system directing the way. For example, if there was a lookout using overhead cameras to tell Bob which way to turn, that would be an example of central control. Even then, the instructions could be quickly dismissed if contrary information was spotted on the ground.

Figures 5.2(a) and (b) summarize enormously detailed processes. Each node in the network graph might represent another whole network that also provides alternative pathways but at that smaller scale. We are free to represent very large systems, like the hypothetical one modeled in Figure 5.3, at just one level in its hierarchy using a low-dimensional state space, as we did in Figure 5.2. Cellular functions may be categorized by scale. For example, metabolism may be broken down into functional metabolic sectors each consisting of metabolic pathways that are themselves composed of chemical reactions [Palsson (2006)].

While coarse-grained models can have enormous descriptive value, their use in addressing detailed questions are limited by their mathematical intractability; for example, there can be hundreds of equations at the molecular reaction level, each with many parameters. There might be fewer equations at the cell level but their parameters are determined by detailed functions that vary significantly with the equilibrium state of the cell. Fine-grained models are used in systems biology to understand cellular mechanisms, whereas coarse-grained models help us visually communicate global network operations. A comprehensive mathematical genome-scale model for how a mammalian cell functions, including all signaling, metabolic, and regulatory functions, does not currently exist and may never exist because

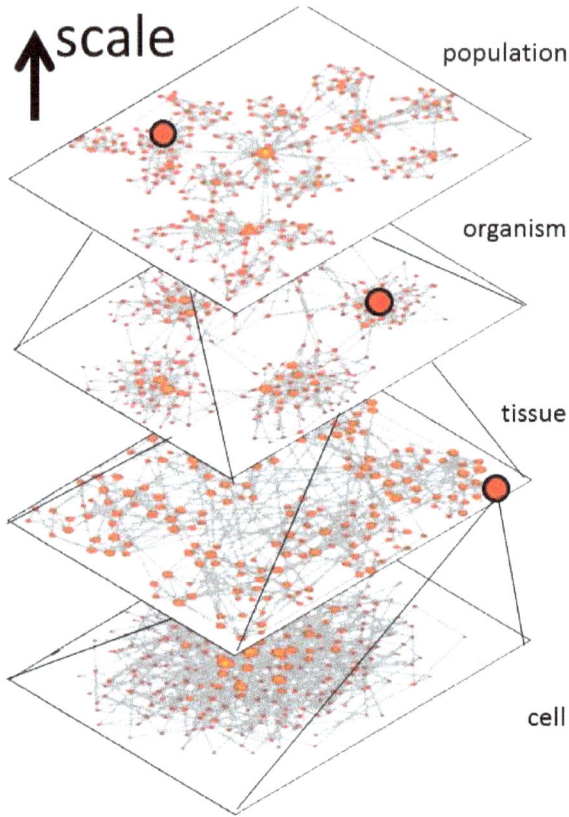

Fig. 5.3   A graph-based representation of a hypothetical biological hierarchy. The highest level is dominated by interactions among organisms in the population and their environment. Within an organism, functions cluster around organs. Organs are collections of different tissue types that are composed of cell types classified by their functional properties. Properties at each level are interconnected broadly (not shown) with properties at other levels. Graphs representing biological systems are not fundamental; they change as required to help researchers understand biological properties.

of how rich the emergent properties become as the dimensionality of the models increases in scale to that of the human genome. It is likely that different models may be required for answering questions about biological systems at different scales, and there may never be a comprehensive model of a mammalian organism. For this reason, generalized systems analysis that operates at all scales could provide the best answers. For example, medical conditions are often most effectively managed on the scale of the organism or population even when the root causes operate at a microscopic level.

Example. Consider human infections by the hepatitis B virus (HBV). Some people have strong natural immune responses to acute HBV exposures that quickly deactivate the virus, but only after inflicting extensive hepatocyte (liver cell) damage that severely impairs liver function. Those with a weaker immune response may have less hepatocyte damage, but retaining the virus is a progression to chronic hepatitis B infection with long-term risk of developing liver cancer (hepatocellular carcinoma). The chance of chronic HBV patients developing liver cancer is 100 times that of the uninfected population, and the 5-year survival of these cancer patients is only 10%. Neither system response can be classified as desirable.

The best protection from HBV today is immunization during infancy. Medicine focuses on immunizations and treatments of individuals and populations – a large-scale system view – because of the importance of minimizing exposure and the central role of the immune system in determining injury and prognosis. Nevertheless, treatments themselves are based on the reductionist science of cellular decision making. Physicians and government officials must keep in mind the human system at all scales as they formulate medical solutions for individuals and make policy-level healthcare decisions for the population. Both can be overwhelming jobs when done well.

This multi-scale view is important for investigating all systems. If one is interested in building a large business selling shirts at concerts, investors will want to see a business model that considers processes related to materials manufacturing, product marketing, shipping and related factors. Of course, business planning might invoke larger concerns about the state of cotton agriculture, texture worker unions, effects from global trade of materials and transportation costs, but these remote issues are likely to have a minor impact on our T-shirt vendor. When modeling these large-scale economic systems, the mechanisms of direct shirt selling at a concert may be just one node within one node ... of a much larger network. As emphasized in Chapter 4, model designers need to carefully focus their attention on those scales impacting the questions at hand, although the model user must remain mindful that larger and smaller scale system properties can create destabilizing conditions influencing predictability. The principal challenge with managing biological systems is the enormous size of each and its strong interdependency on many other systems.

Boomer Bob arrives at the vendor table to find a great all-cotton T-shirt exactly as he hoped it would be. As he winds his way back to his seat in time for the next music set, Bob reflects that the model in Figure 5.1(a) on which he based his strategy was pretty simple and yet effective in this situation. To declare the model as generally valid, however, Bob would need to repeat this activity at many concerts many times to experience more realizations of the variable properties that influence acquisition time. If he thought about it a little longer, Bob would also be quite satisfied that his cellular metabolism kept up energy production in his 60-year-old body allowing him to enjoy his musical adventure.

5.2   Cellular Complexity

Biological systems reveal extraordinarily high levels of hierarchical complexity, as hinted at briefly above and in Chapter 1. Let's explore the complexity of the human system told from the perspective of a cell. Individual multicellular organisms begin existence as one very special cell – a zygote or fertilized egg. Following birth and two decades of healthy development, this cell will have divided repeatedly to form an adult human composed of roughly 80 trillion native cells, many with highly-specialized forms and functions (cell differentiation) while others remain undifferentiated and ready to adapt as needed. Every human cell in your body that has a nucleus contains the same DNA instructions as that first cell.

There are 23 pairs of chromosomes in each human nuclei, half of each pair was contributed by one parent. Chromosomes are information-bearing, long, nucleic-acid sequences that are mixed with proteins to keep the chains compact, organized and accessible when their information is read. Nucleic acids are long polymeric chains whose links are nucleotide monomers. Each nucleotide has three parts: a deoxyribose sugar, one or more phosphate group, and one of four possible nucleobases; the first two parts offer structural integrity and energy sources, and the third part is a "bit" of genetic information. One of four nucleobases attached to the sugar-phosphate backbone in DNA form the "letters" of the genetic message. Two of the nucleobases are purines, adenine (A) and guanine (G), and two are pyrimidines, thymine (T) and cytosine (C). Nucleotide strands covalently bond sequentially as links in a nucleic-acid chain to preserve the order of the nucleobase sequence. Each long nucleic-acid strand is paired with its complementary strand through weak hydrogen bonds. Once formed, the double

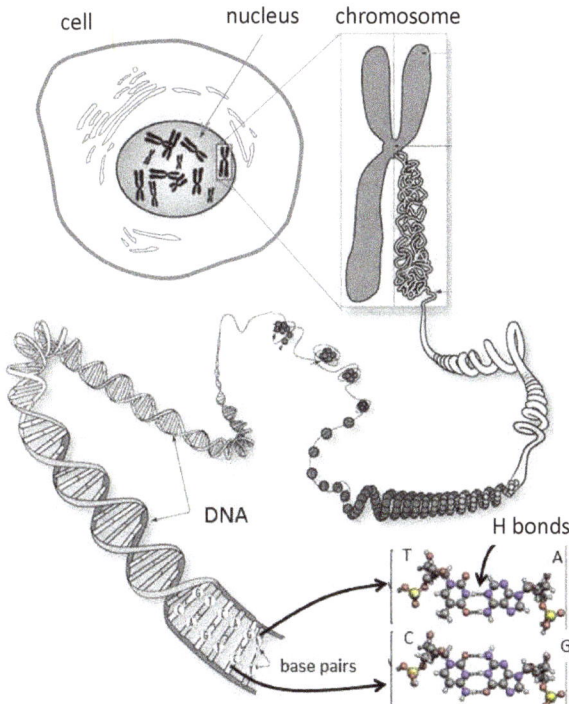

Fig. 5.4 Chromosomes in a human-cell nucleus are a combination of DNA molecules mixed with proteins. At the smallest scale (bottom right) are sequences of base pairs consisting of T nucleotides joined with A nucleotides and C joined with G. Double stranded nucleotides form a helix that wraps itself around histone proteins. In this way, meters of DNA sequences are efficiently packaged into chromosomes to appear somewhat like 20th century telephone cords. Chromosomes easily fit into a 10-micrometer nuclear diameter while keeping their information accessible for copying. Based on several figures in [Lodish *et al.* (2004)].

stranded DNA molecule twists to form a helix that folds into the chromosome shape illustrated in Figure 5.4. All the information about constructing your human form and biological functions is recorded in those nucleic-acid sequences. If you were to unfold the meters of DNA curled up in your chromosomes, you would find there are more than 3 billion base pairs formed from the 4-letter molecular alphabet A, T, C, G.

Cells are very efficient at packaging long DNA strands, and hence they carry extraordinary amounts of information in a tiny space. Genetic information is organized in a complex way, not unlike that of a book (a common descriptive metaphor). A book offers long sequences of about 100 distinct characters. The shortest sequences form words that convey small ideas. Sequencing words into sentences, sentences into paragraphs, and so on to larger scales, introduces an information hierarchy in which a very complex range of ideas can be expressed. If nucleotides are analogous to letters, then genes might be analogous to words, sentences and paragraphs. Like words, some simple genes convey specific information about an organism's characteristics. But that analogy is too simple. Also like well-placed words, DNA segments gracefully harmonize information to form sentences and paragraphs that provide a rich palette of traits, like arm length and blood type, which we refer to as *phenotype*. The nucleotide segments corresponding to some observable output like a protein is rather complicated to define because genes link to other genes as a way to regulate protein expression (this is the *operon* concept in genetics). Similarly, it can be hard to identify exactly which words in a paragraph correspond to each concept. Regulatory networks operate like adverbs and adjectives do when linked to verbs and nouns in sentences, to emphasize and moderate gene expression. For example, "The explosion was earth-shattering! Its blast flung me backwards hard against the wall..." is more detailed than "There was an explosion. I fell back..." They covey the same facts, but the former adds important new meaning.

Collectively, genes archive the history of the accumulated successes of life on Earth for easy and regular reference by cells. However, proteins are the currency of the cell, like money in society. Proteins must first be constructed before cells can carry out functions including *gene expression*. The most basic function of DNA is gene expression, which is a process frequently resulting in the assembly of RNA and protein molecules. RNA is another family of nucleic-acid molecules, but this one acts as intermediary between the DNA instruction and protein building. If DNA is a long-term information-storage device, RNA is for information reading and short-term storage.

$$\text{DNA} \xrightarrow{\text{transcription}} \text{RNA} \xrightarrow{\text{translation}} \text{protein}$$

Fig. 5.5   Process leading to protein generation in a eukaryotic cell.

At its simplest, as illustrated Figure 5.5, gene expression occurs in two stages. The first stage is *transcription* where one form of RNA is guided by proteins called transcription factors to activate a segment of the DNA molecule. Like finding a book on how to build furniture, opening it and reading a passage, gene activation includes unfolding a chromosomal segment and copying a specific DNA segment into RNA. The second stage is *translation*, where information gleaned from reading the book guides the construction of furniture. In the cell, this occurs when RNA links with a large molecular machine called a *ribosome* that is composed of RNA and protein molecules. Reading the RNA template, ribosomes select from among 20 amino acids found within the cell cytoplasm to build amino acid chains called *polypeptides*. These polypeptides then fold into very specific proteins shapes that have specific functions. Protein information is encoded in both the amino-acid sequence and its 3-D shape. Some proteins, e.g., collagen and elastin, offer support structure, while others provide intercellular communications, enzymes accelerate chemical reactions, actuation fibers facilitate cell movement, and so on.

Gene expression is how *genotype* (the genetic code of an organism) generates *phenotype* (organism characteristics related to structure and function). Protein-coding DNA specifies at least 20,000 human genes, but those genes are found in less than 2% of the DNA molecule. The other 98% performs a variety of functions including regulation of gene expression, pre-translational RNA editing, and numerous other known and unknown legacy functions. Your DNA carries all the inheritable biological information there is about you, including how to build all the molecular machines your cells require to be healthy and active, as well as information about millions of years of ancestral successes. As with any history, DNA is a story told by contest winners – survivors.

The cellular machine is Nature's most extraordinarily sophisticated invention. Each is capable of acquiring the materials and generating the power to run factories building whatever RNA and proteins are needed to run a multicellular organism. They do this by carefully selecting information from a constant blizzard of microenvironmental signals bathing most cells in our body all the time. Cells are loaded with protein-based receptor molecules acting as locks waiting for signaling-protein keys that might float by with instructions on what the body wants the cell to do next. Signals can be sent by cells in the body and from substances absorbed into the body from the environment. Most cells can divide to replace themselves at some programmed or sensed rates as they are worn out or damaged. For example, cells lining the gastrointestinal tract are replaced every few days while cells in the heart and brain may be with us our whole lives. Some cells fade into inactivity as we age, even though they remain alive (cellular senescence), and others politely self terminate (apoptosis) once they sense they are no longer needed or may be harmful.

Cellular genomes employ several highly sophisticated parallel networks to manage all of their onboard manufacturing and communications facilities. These include metabolic networks for energy production and resources management, and signaling networks where cells listen to other cells in the organism and transmit this information internally to their nucleus along protein cascades. There are also protein-based regulatory networks that edit signal messages by (a) turning genes on and off or (b) modifying the intensity of cell responses by issuing enzymes (a form of protein) that regulate the speed of reactions. It is truly amazing that trillions of little sensory-based, computational decision makers that compose our bodies, many forming and dying at different rates, are able to smoothly coordinate their individual efforts over a lifetime. Coordinated biological functions issue from all cells in the body like music in a hall from an exceptionally-large orchestra, as long as each component is well trained and listens carefully when responding to output from other components and emergent properties of the whole system.

Now consider that there are 10 microbes living on and in our bodies for each human cell, many of which we need to stay alive (see Fig. 5.6). Each of these foreign cells has its own distinct DNA-guided machines, generating proteins that add their distinct voice to the communication symphony of our native cells. This inter-kingdom cooperation, termed the *human microbiome*, is essential for healthy efficient human digestion, metabolism, and appropriately-tuned immunological surveillance. If we're healthy, these hundreds of trillions of cells, both foreign and domestic and each with sophisticated sets of internal networks, synchronize to create a homeostatic equilibrium state that remains adaptable enough to negotiate variable environments. To study multicellular life, especially if one considers the breathtaking capabilities of the human brain and its senses, is to confront systems of mind-boggling size and complexity.

Make no mistake, to acquire the deeper understanding of biological systems necessary to advance medicine, we must learn to apply engineering sciences on a truly remarkable scale! Training in condensed matter physics, organic chemistry, psychology, medicine, environmental sciences, sociology, mathematics and other fields are all necessary for studying biological systems at different scales. It took the revolutionary ideas flowing from the

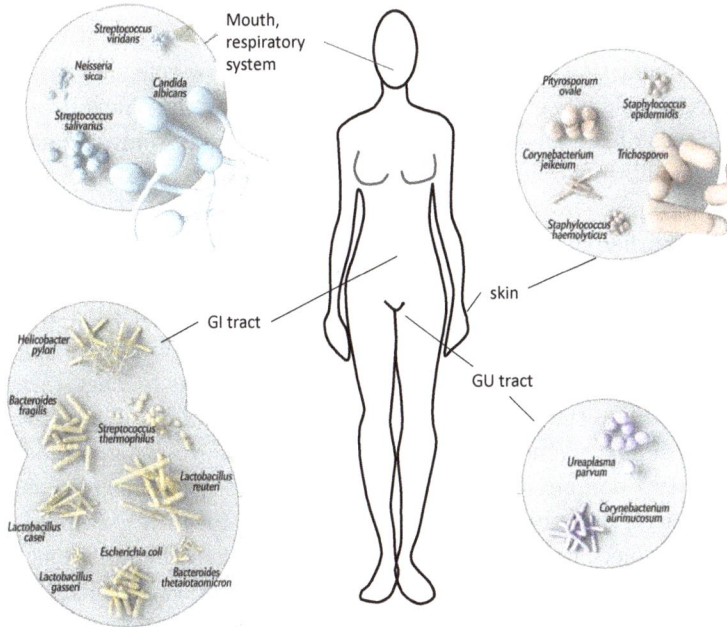

Fig. 5.6 Illustration of the human microbiome based on the description in [Ackerman (2012)].

amazing technology developments that built the field of molecular biology during the last 50 years for scientists and engineers to begin to understand the importance of system engineering to the existence and stainability of life. It seems to us that new systems engineering tools must be developed to analyze the true extent of the microbiome on human health.

Life may be the most complex system we know. Even as we continue to discover the engineering details of life processes, we are challenged by the task of making sense of it all. One hope for success might be that all complex systems share general principles governing their behavior. Then lessons learned from simpler systems, like the mechanical systems of Chapter 4, might guide the development of methods for analyzing biological systems. A challenge that scientists and engineers have yet to master is how best to navigate the massive hierarchy of biology. The typical scientific approach to the task of analyzing complex structures is to study the fundamental unit. In the case of biology, that unit is the cell, the smallest structure that can claim to be alive.

5.3   Stability

Before summarizing one of the many theories on how cells may have come
to be, we review two essential properties of systems that have become the
cornerstones of this book. Those properties are (a) systems can exist in a
state of equilibrium, and (b) systems have a tendency to seek and remain in
an equilibrium state. Equilibrium implies some type of *stability*, although
we saw in earlier chapters that unstable systems about to undergo a transition are in a state of equilibrium until the transition begins. Life seems
to exist in a highly dynamic state of equilibrium that can seem always on
the verge of collapse but remains stable nonetheless over a lifetime. We
discuss in Chapter 6 that not only do complex systems manage variability
to remain stable, they require variability to maintain stability. Let's first
consider a simple example of chemical stability from your high-school days.

Place two sodium atoms (2Na) near one *covalently-bonded* chlorine
molecule (Cl_2) and the different atoms combine to form two more-stable
compound molecules called sodium chloride (2NaCl). The different atoms
join through the process of *ionic bonding*, where two neutral sodium
atoms donate one orbital electron each to become sodium *ions*. Ions are
atoms with a net charge such as $2Na^+$, where + indicates each atom is
missing an electron. Also a chlorine molecule separates its two atoms as
each accepts an electron to form two chloride ions ($2Cl^-$, where − indicates each atom has acquired an additional electron). If there are many
such atoms around, then bonded pairs form and arrange themselves in regular *crystalline* patterns. This occurs passively because ionic bonding does
not require additional energy, in fact the process releases some of the energy
stored within the atomic systems as heat. Consequently, the NaCl product is more *stable* (has lower equilibrium energy) than its precursor forms.
Stable forms of molecules are more prevalent in nature, as you might guess.

The organic molecules of living systems are also bonded together in
ways that achieve the stability of low-energy states. The atoms within large
DNA molecules (Fig. 5.4) are of similar size. Therefore, covalent bonds are
formed instead of the stronger ionic bonds. Although organic molecules are
more fragile than salts, they nevertheless form durable polymer chains that
preserve the information found in nucleotide order. The two nucleic-acid
strands composing human DNA are joined by hydrogen bonds at complementary nucleobases: A is matched with T and G is matched with C as
shown in Figure 5.4. Hydrogen bonds are about 10 times weaker than
the covalent bonds holding the sugar-phosphate backbone together, so the

double strands can be safely separated (unzipped) when they need to be copied but each strand is normally very durable. Unzipped DNA sequences are copied to RNA molecules that have only one strand formed from the complement of the DNA segment it is copying. Chemical stability is an essential feature of all living cells.

5.4 A Story About the Origin of Cells

The basic unit of human life is the cell and the essence of a cell is to store, communicate, and express the genomic information recorded in DNA. The most basic function of a cell is to replicate DNA in different ways: It does so to (a) read segments that express a cellular function, (b) make new cells from a copy of the entire genome, or (c) make gamete cells that combine in sexual reproduction to form a new organism. So it is reasonable to assume that cells arose from some manner of precursor *replicator molecule*. Replicator molecules have an ability to remain stable long enough to copy themselves regularly and with high fidelity. There are several versions of the replicator-molecule story leading to cell formation that can be found in the literature. A very compelling and readable story is told by Richard Dawkins in *The Selfish Gene*, [Dawkins (1989)]. We highly recommend Dawkin's book, and we'll try to explain why.

The first replicator molecules are thought to be an unusual form of RNA. The *central dogma of molecular biology*, at its most basic, tells us that cellular DNA generates RNA molecules that generate proteins in a cell (Fig. 5.5). To complete the cycle, proteins acting as enzymes catalyze the reactions needed to form new DNA. So it seemed for a long time that none of the three structures, DNA, RNA, or proteins, could be the first replicator molecule because each needed another to complete replication. It's the old chicken and egg conundrum.

Then scientists [Zaug and Cech (1986)] found it was possible for some RNA molecules to fold and function like proteins. Although multifunctional RNA molecules are unusual, their discovery showed that RNA could auto-catalyze reactions that enabled self-copying – they could disassemble molecules and then use themselves as a template to reassemble the parts that replicated their form. Rudimentary replication was possible.

There is evidence that a billion years ago the Earth had plenty of organic substances lying around. These could have included nucleic and amino acids, perhaps some from extraterrestrial origins delivered on asteroids and comets. Nowadays bacteria would gobble up unprotected organic substances, but we are talking about a time on Earth before life existed. It is not too hard to imagine that, as countless numbers of organic molecular fragments came into contact over millions of years, by chance some formed a stable replicator molecule, perhaps something like one of the folding RNA molecule described above. Once a replicator existed, it was able to make copies of itself from the raw materials in its surroundings.

If the energies required to form the most common organic molecules were about the same, then all molecular forms would be found in nature at about the same concentration. We can then say that molecular systems of primordial Earth were *disordered*. A replicator molecule changes that situation by increasing its numbers at the expense of other forms. By reproducing, the replicator managed to generate order within a disordered environment, and the preference for one molecular form over others adds *information* to its molecular system. Hence, *replicators were information generating machines.*

The replicator brought a type of stability into existence unlike anything before it. Replication followed the laws of physics like other chemical reactions, but something was different. Replicator molecules were more fragile than other stable forms of matter, yet these fragile nucleic acid structures prevailed by actively replicating many copies before the original molecule fell apart. Its molecular form dominated the molecular system if its copies were accurately formed and remained stable long enough to make more copies.

Copying *errors* would certainly not be of help to the replicator making them because reduced copy fidelity is a loss of the original form; a loss of information. If large errors were made, the copies were unlikely to be properly-functioning replicators and so the original form was not preserved. Any random error, large or small, is most likely to introduce a weaker copy that quickly fades, and so the individual replicators that were most successful were those that made the fewest errors.

Some errors were small, so the product was another functional replicator. Rarely, some of the altered copies were more prolific or longer lasting than the original. Those copies became strong competitors for the limited molecular resources, and limited resources would further reduce the stability of the original form. It must have been very rare, but some errors helped replicators diversify their form by introducing an innovative feature that provided a survival advantage. Stronger offspring would quickly dominate the local molecular ecosystem once they appeared. So copy errors that are always bad for individuals making the errors became critically important for long-term stability of all replicators trying to survive in a changing environment. Today copy errors in DNA, the replicator's successor, provide diversity that increases stability for the population at the risk of the individual to drive evolution. This is an example of variability promoting stability.

Less stable forms of replicators on average generated fewer copies that preserved the original form. More stable replicators made fewer copy errors and might have lasted a little longer to populate their environment more densely, at least until the raw materials for copying became scarce. Since the copying process implies an ability to break up organic matter (catabolism) before reassembly (anabolism) into the preferred form, there would be little difference between raw materials and other replicators. Dawkins suggests that replicators evolved methods of protecting themselves from hungry neighbors.

One form of self-protection may have been for the replicator to wrap itself in a protective layer that was somewhat impervious to the catabolic effects of its neighbors and the hostile environment, particularly solar radiation. Just a little protection would provide an enormous advantage to that newly evolved form, just as a tiny bit of vital financial information made available to a few investors can make them vast profits in the stock market at the expense of other investors. Of course, for any revolutionary change in replicator form to not be a passing fancy, the instructions for building the protective layer (eventually a *cell membrane*) must be stored in the nucleic acid sequence of the replicator itself so it can be passed on and reproduced in future copies.

If this scenario is even close to being true, then it could describe the formation of the first cell-like structure. The proto-cell was composed of a membrane permeable to resources but impermeable to external toxins. Within the membrane, the replicator molecule included instructions on how to build protected copies of itself. Replication was no longer just the process of copying a molecule. The improved replicator carried instructions for building a protective membrane, which was a *proto-gene*. Dawkins calls the membrane innovation the replicator's *survival machine*.

Of course, the membrane and gene innovations were huge simultaneous steps that likely required an accumulation of many smaller changes, perhaps over thousands of years, before anything reliable could be passed along. Because the record of molecular-system innovations was not preserved as it was for fossilized organisms that followed, we can only guess about the sequence of events leading to the first cell.

Conjectures about what might have transpired along the way to produce the first cell are acceptable provided they are consistent with all known facts about what actually occurred. A *conjecture* becomes a *theory* when it is the story most consistent with the facts. *Theory* becomes *law* once it has been extensively tested and found consistent with all known facts. *Conjecture* become *myth* when inconsistencies persist. A scientific theory is nowhere as wishy-washy as the general public frequently assumes when making statements like "It is just a theory". More accurately, they should say of a conjecture that flies in the face of facts, "It is just a myth". Theory is science's tool for approaching the unknown to decide if it is unknowable, and for guiding experimental searches in case it is knowable. Thus conjectures on the origin of the first cell must be periodically reviewed as new facts become available.

5.5   When Did Replicators Become Living Cells?

The proto-cell described above is a replicator molecule contained within a self-made membranous survival machine. But is it alive? Life has been defined in many different ways but always in terms of observable properties. *Fundamentally, a living cell is a permeable space distinct from its environment that contains a record of its form and function and an ability to use energy, replicate, and to sense, respond and adapt to its environment.* The proto-cell discussed above has these characteristics except an ability to directly sense and respond to the environment. Breaking down and assembling organic substances to replicate would have originally required energy supplied by the environment until an innovation appeared that allowed energy to be produced within the membrane.[1]

Replication errors enabled adaptation from the beginning so long as the errors were small and the error rate was low. If a replicator error provided even a small benefit and it persisted, then the error became an indirect way of sensing and responding to the environment. This form of sensing is indirect because the response would appear in copies but not necessarily in the individual to which the stimulus was applied. It might be said the proto-cell became alive once it developed reliable internal sources of energy production and direct sensory responses.

---

[1]There are multiple theories on the origin of the energy-producing *mitochondria* found in the eukaryotic cells of plants and animals. One theory is that mitochondria were believed to be once independent prokaryotic cells (bacteria) that became incorporated via endosymbiosis into eukaryotes to supply an internal energy source. As in business today, that which is needed but not developed internally must be acquired through synergistic mergers.

A replicator form endured only if many copies of that nucleic-acid sequence were accurately produced during that individual's existence. With a small error rate, any hard-won chance innovations born of errors were preserved within the template molecule to accumulate and develop over time. If the error rates were too low, innovations would appear too rarely to make a difference and "replicators" as a form would stagnate and disappear under environmental challenges. If the error rates were too high, innovations could not be reliably recorded so they could accumulate. To exist long enough to reproduce within a limited-resource environment, a structure must sense environmental information (Where are the resources? Where are the hungry neighbors?) and respond in ways that promote survival of the form.

Information must flow from the replicator/cell for life to persist. Cells are tiny *anti-entropy (structure building) machines*. Cell systems thrive in resource-rich environments that provide the means to build order from disorder, which is likely why some evolved to exploit the advantages of multicellular organisms. We consist of trillions of cells because that is a stable way to accurately record and pass along the instructions of past evolutionary successes. What makes life unique among other forms of mass-energy is its ability to acquire, build and traffic in information as it adapts to a changing environment.

Let's explore what is meant by "information" given our position that it plays a central role in defining cell-system properties. There are rigorous mathematical definitions of information that we avoid in favor of building general reader intuition.

5.6 Information

Information is a rather nonspecific word applied to technical and everyday ideas. For our purposes, information is a property of *signals*, which are any sequence of events. Information quantifies how much listener uncertainty can be reduced by receiving the signal, and its value is related to the knowledge gained. If I'm wondering about the current time and someone correctly yells "3 o'clock," the sequence of sounds I hear provide me with information because that signal reduces my uncertainty and increases my knowledge of the current time, assuming I understand the language of the signal.

For a more quantitative example, assume a listener is trying to predict the next number in a sequence, as one hopes to do when playing the lottery.

If all possible numbers are equally likely, then the history of lottery numbers selected over the past few years is uniformly random and time independent. By design, lottery histories must contain no information that might help ticket buyers select numbers. Hence our uncertainty about the next number in the sequence is unchanged by simply listening to past winning numbers. However, if the likelihood of event occurrences in the number sequence deviates from uniformly random in a measurable way, that sequence can be said to contain some amount of information. We suggested above that the replicator molecule was a device that increased the information within a system of organic molecules because it drew those materials from a random collection to build copies of itself.

An example similar to lottery number selection is predicting outcomes of coin tosses from a past history of tosses. Coins that are uniformly random or "fair" display on average a "heads (H)" result 50% of the time and a "tails (T)" result 50% of the time. To be a probability statement, the percent occurrences of all possible alternatives, in this H and T, must add to 100%. We gain no information about future tosses from a history of fair-coin toss results because the sequence does not reduce our uncertainty about the next result. If I suspect a coin is slightly "unfair," specifically, if the probability of a heads is 49% and the probability of a tails is 51%, the information situation changes. Observing a sequence of events, we can estimate information from *statistical bias*. In this case, bias is the difference between the average numbers of T results that we actually measure and the number expected if the coin was fair. Flipping the coin 1000 times, the percent bias equals $(510-500)/1000 = (51-50)/100 = 1\%$. If we are betting on these results, we should guess T each time to exploit the information from the 1% bias. Even if we are unaware of the bias, the information exists in the sequence and can be measured retrospectively. Scientific theories that predict unknown system properties are hinting that new information might be available. It is this promise of new knowledge that motivates many experimental analyses.

Computational machines evolved from the idea that information can be encoded in binary number sequences like H, T or 1, 0 or on, off. Binary sequences are now widely used to communicate information between locations, extract information from data sequences, etc. The sequence itself is not enough for the information to have value. All parties must agree upon (or work to discover) the language used to read and write the sequences of binary events (bits). From right to left, electronic bit strings can be read by breaking them into segments of eight *bits* to form a *byte* of data. One byte encodes $2^8 = 256$ distinct states that may represent alphanumeric characters on a keyboard, color intensity in video data, and other signals. For example, if you type a lowercase "m" on a computer keyboard, the keystroke is converted into the binary sequence 01101101 and sent as electrical pulses to a receiver. If the listener understands the language is ASCII code,[2] the bit string is understood to represent letter "m". Electronic signal transmission always adds random errors to the message (noise) that reduce the information content of the message unless the errors are corrected. Noise cannot generate new information that a listener would be prepared to read and understand; "innovations" born from transmission errors are all *noise*.

Nature uses chemistry to store the instructional information for human life in nucleic acid sequences. The genetic signal is transmitted to new cells as a self-starting (bootstrap) instruction set for how to build, operate, and reproduce a multicellular organism. In effect, organic chemistry is the universal language of DNA information transfer. DNA contains a digital code, composed of the letter-based quadrary sequence A, T, C, G instead of the two-number binary sequence 1, 0 of digital computers. Unlike the digital computing world, the chemical world of DNA pairs its digital sequences as shown in Figure 5.7, always A with T and G with C. Complementary matching of base pairs stabilizes the molecule for long-term storage while making it easy for the cell to unzip the hydrogen bonds connecting base pairs when it is time to read and copy the code; after all, both strands carry the same information.

In the digital computing world, 8 bits form a *byte* of information that codes for the alphanumeric characters of language, among other things. In the chemical world of protein-coding DNA, three base pairs form a *codon* of information that specifies the amino-acid sequences of a protein among other things. A four-letter alphabet grouped three at a time can specify $4^3 = 64$ amino acids. However, there are only 20 amino acids and some start-stop punctuation-like signals. We say amino-acid codes are degenerate because each amino acid can be specified by more than one codon. *Degenerate coding* was avoided in the ASCII lexicon.

---

[2] ACSII is an acronym for American Standard Code for Information Interchange. Since there are roughly 100 characters on English language keyboards, 7 bits (two-letter alphabet in groups of seven gives $2^7 = 127$ distinct states) is usually enough for human-computer interfacing. However, industry standardization of the 8-bit byte extended ASCII to 8 bits. If the leftmost eighth bit is unused, it is set to zero as we did above.

ATCGATTGAGCTCTAGCG
|||||||||||||||||
TAGCTAACTCGAGATCGC

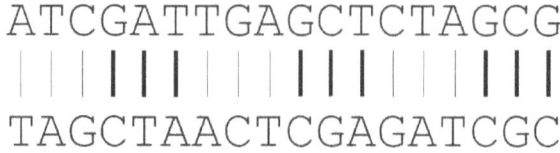

Fig. 5.7   An 18 base-pair sequence of DNA is shown. Every three elements in the sequences define a codon. An unzipped section of a DNA molecule is copied into a messenger RNA sequence (not shown) during transcription before being translated into the amino-acid sequence of the protein to be built. Codons specify the order of amino acid assembly as messenger RNA is read by transfer RNA in a ribosome during the translation phase of gene expression.

The long-term storage of information in DNA molecules is transcribed into the short-term information storage of RNA molecules using a slightly modified nucleic-acid alphabet. RNA information is then translated into protein sequences that perform cell functions if they are able to correctly fold. Protein information is encoded into 3-D molecular structures as well as the amino-acid sequences.

We mentioned that some proteins remain in cells to serve as enzymes. Enzymes catalyze chemical reactions and regulate metabolism by controlling reaction rates. Other proteins found inside and outside the cell and in the membrane are used for ligand[3] binding and cell-cell signaling. Examples of ligand-binding proteins include antibodies that bind antigens (foreign substances) to neutralize their systemic effects in cells, and hemoglobin in red-blood cells that bind oxygen for metabolic transport. There are at least 1300 different enzymes in cells that are constantly being formed and broken down so the amino acids can be reused. Fibrous proteins, like collagen and elastin, are polymerized into different shapes to form the connective tissues that provide overall structural support for cells in the body. Inside most cells, there is a protein cytoskeleton that enables cells to move within the extracellular collagen-protein matrix. Collagen and other associated proteins form tissue stroma that gives the body its mechanical properties. Stromal stiffness is sensed by cells to control cell proliferation rates, phenotype, and other essential functions. Each component of a cell encodes information vital for the functions of life. Hence, cells are *information machines* in the sense that they are constantly building the molecular structures that support life.

---

[3]In this context, a ligand is a molecule that binds to a cell's receptor molecule. This is often a first step in cell signaling.

5.7   Summary

The fundamental unit of life is a cell. Living cells are computational systems that receive signal information from their environment. They process that information according to stored genomic instructions and then decide which protein machines to build that can respond appropriately to the signals received. Cells have their own internal factories that can implement their genetic instructions, including resource management (materials selection and transport), communications, and self-regulating abilities to hibernate, reproduce, or self-destruct.

We find it convenient and meaningful to model cell systems as a set of parallel interconnected networks, separated according to what we reason appear to be separate functions. These networks respond to environmental signals and cooperate with genomic instructions to self regulate and achieve an equilibrium state. The separation of the cell system into metabolic, signaling and regulatory networks is artificial because they cannot operate independently. Nevertheless, we tend to model separate networks because it helps us organize the many parallel cell functions. Remember, models are built to understand and predict the behavior of systems. If they are successful at their job, they are useful even if they have obvious limitations in other respects.

Network models can be scaled up to describe how cells form tissues and organs within multicellular organisms, and how populations of organisms form societies that interact among themselves and with other species in the biosphere. At every spatial scale, down to the cell, living systems must continuously traffic in information to remain alive, which requires tremendous amounts of energy. The exchange of information from the use of chemical, mechanical, and electrical energy makes life different from other forms of matter, as we discuss in more detail below. Life today is a surprising property that emerged from the adaptive chemistry of millions of years of evolution applied to replicator survival machines.

Advances in genetic research since the Dawkin's book originally published in 1976 have prompted a new book called *The Society of Genes* [Yanai and Lercher (2016)]. The latter authors note that while some genes do behave selfishly, others cooperate in complicated ways to form gene systems that resemble behavior of individuals and groups in society. They give examples to show genes are not simply an instruction set. There are genes that only copy and insert themselves throughout the DNA molecule. Some do little or nothing that we can discern while others operate extensively to perform large-scale operations throughout an organism. Some genes are dedicated to mundane tasks and others play a management role for large groups of otherwise unrelated genes. Some genes misbehave prompting others to fix their messes. Indeed, there is evidence to suggest genes form a complex system – a society – that expresses a range of emergent properties that we are still discovering.

5.8   Eigen-analysis

As information networks, cells may be understood from their eigensystems that are subject to the governing principles of all complex systems. To make effective use of eigen-analysis, we must interpret cell properties in terms of eigenstates. We approach this interpretation task through analogy with a previous analysis of the mechanical properties of a wooden rod.

Assume a wooden rod is held in our hands near either end, as in Figure 2.1. For an input bending force pattern applied to the rod system, we can measure its response as output bending deformation to describe system properties. Observable properties of the rod system are found by quantifying the relationship between input-output measurements. There are many other observable mechanical properties; e.g., those describing twisting, compression, and tension-property deformations, as we already described. Eigenfunctions of the rod system are uncoupled representations of those observable mechanical properties. We can activate individual eigenfunctions experimentally by identifying through modeling the input force patterns that isolate each deformation mode of the rod system.

Rod-system eigenvalues depend on material and geometric properties of the wood. Large eigenvalues indicate a large external energy must be supplied to activate its eigenfunctions. Essentially a stiff wooden rod has large eigenvalues representing its mechanical properties. If holes are drilled through the rod, its eigenstates change according to changes in geometric properties. Now, when the same force is applied, the expected response is a larger deformation indicating the holes acted to reduce the eigenvalues. If an eigenvalue for a rod goes to exactly zero, then any force applied along that eigenfunction will initially move the rod along that deformation mode as the rod system transitions to a new "broken" equilibrium state.

By analogy we can extend this analysis to cell systems. Consider the input-output relationship for the highly-simplified network model of Figure 5.2(c). Assume this model represents a cell network. The input is a signaling ligand entering the locale from outside the cell to bind to a receptor

protein embedded in the cell membrane. As the first step in the signaling process, ligand binding triggers a chemical transformation within the
receptor protein that, in our highly simplified model, initiates a sequence
of seven more chemical transformations, eventually resulting in a protein-
production output. In real cells, signaling cascades involve a sequence of
hundreds of chemical transformations, including several membrane-receptor
activation processes, internal signaling-protein cascades involving enzyme
cofactor exchanges, all the mechanisms of gene expression, energy production and transport activities. In real cells, as in our simple seven-step model,
we have ligand input to the cell signaling network system generating protein
output in accordance with cell properties.

Each of the seven chemical transformations (fluxes) in the sequence of
seven may be represented by one black-line edge in Figure 5.2(c). The
corresponding intermediate chemical substrates are represented by nodes
along the path connecting input to output. The input-output relationship
for this sequence is an observable property of the cell's signaling network.
Each observable property can involve the activation of multiple network
paths, and, knowing something about all possible paths enables us to find
uncoupled cell properties from network eigenvectors.[4] The cell-biology literature refers to *cell pathways* that might give novice readers an impression
there are hardwired paths through cell circuitry. This is not the situation.

The simple network model of Figure 5.2(c) shows that other reaction
paths connect input to output (gray-line edges) although these might be
less active contributors. We should consider all paths simultaneously as
resources at the disposal of the cell for responding to input stimuli. Eigenvalues of signaling networks measure the ability of a cell to achieve its
programmed response to the input stimuli. A large eigenvalue for a cell
network indicates that little external information must be supplied by the
environment for the cell to achieve its goal. Large eigenvalues for a cell
network indicate the cell has great capacity for achieving its intended response, just as large eigenvalues for a wooden rod indicate the rod system
has a high capacity for resisting deformation, the desired response. The
two systems are compared in Table I.

Systems with large eigenvalues have more diverse resources available to
them for achieving the programmed goal. In the case of a wooden rod, there
are many tightly-bound microfibrils that fortify the mechanical properties

---

[4] *Eigenfunctions* were used to represent rod systems because wood may be considered
a spatially continuous object. We use *eigenvectors* to represent cell networks that are
modeled as discrete components.

with an ability to resist deformation when subjected to input forces. No subset of microfibrils within the wood solely determines the mechanical properties. Similarly, in the case of a living cell, large eigenvalues indicate there is a diverse network of flux reactions and molecular substrates on which the cell can depend when responding to input. A system with few links between input and output has smaller eigenvalues than a system with a rich network of pathways, regardless of how active the sparse connections might be. If one of the network eigenvalues approaches zero, then something fundamental is lost in the cell's ability to respond when a ligand binds with the appropriate affinity to a cell receptor. We describe that "something" in following sections.

Table I. Interpretation of eigenvalues in example mechanical and cell systems

Wooden Rod Support System	Cell Signaling Network System
Desired system output is less rod deformation.	Desired system output is protein production.
Eigenvalues measure the rod's ability to achieve the desired output; larger amount of external *energy* must be provided to deform a rod with larger eigenvalues.	Eigenvalues measure the network's ability to achieve the desired output; smaller amounts of external *information* are required to generate proteins in a system with larger eigenvalues.
One or more eigenvalues → 0 indicates the rod system has modes requiring no additional forces to become active. The desired output is impossible for those modes; the onset of a transition to a new equilibrium state becomes inevitable. Post transition, the system has different mechanical properties and eigenstates.	One or more eigenvalues → 0 indicates the cell network has response modes with no ability to generate DNA-programmed proteins. The desired outcome is impossible through those modes, and transition to new equilibrium state is inevitable. Post transition, the cell has new protein production properties and eigenstates.

5.9   Cell Network Analysis: Example 1

To illustrate eigen-analysis applied to cells, let's examine the network segment diagrammed in Figure 5.8(a). This network was proposed by Christophe Schilling and colleagues [Schilling *et al.* (2000)] as a numerical example of how cellular metabolic networks may be characterized. Figure 5.8(a) is a graph of a sequence of ten chemical reactions labeled by reaction-flux numbers 1 through 10 that can activate five molecules labeled A through E. Signaling events initiated at one of the four input fluxes propagate through the network to appear at outputs via internal flux paths (gray box in (a) defines the system). Because the system is in equilibrium, concentrations of the five molecular activation states do not change with time until we purposefully modify the system.

The input-output reaction fluxes labeled #1,3,9,10 extend outside the cell system while fluxes #2,4,5,6,7,8 are internal. Each reaction is

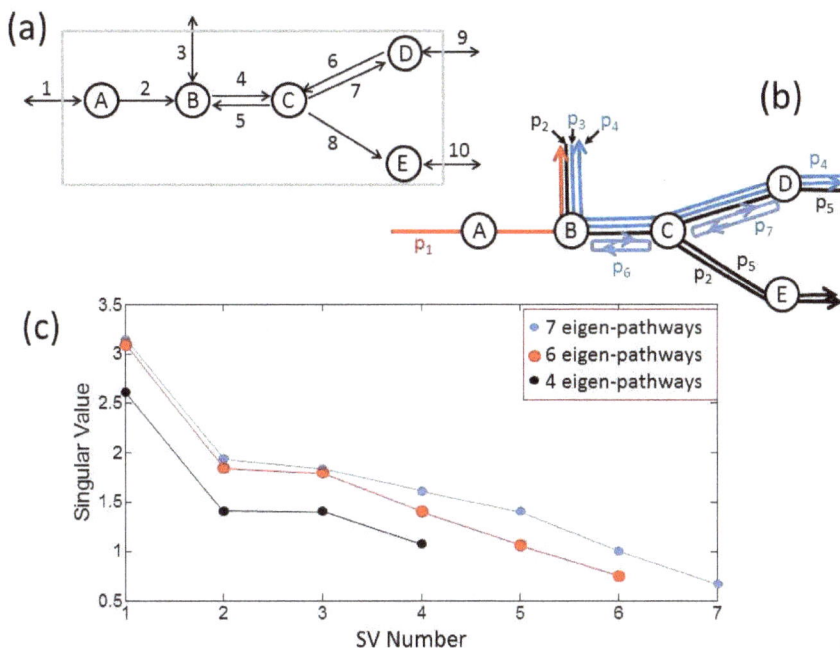

Fig. 5.8 (a) A simple network with five nodes (A-E) and ten fluxes (arrows labeled 1-10; there are 4 input-output fluxes and 6 internal fluxes). (b) Convex analysis applied to the network yields seven generating vectors $p_1, p_2 \ldots p_7$. SVD analysis applied to the generating vectors yields seven singular values, one for each of the generating vectors. Singular values are shown in descending order by the curve with 7 points in (c). Setting flux #2 from (a) to zero eliminates pathway p_1 in (b). The singular values for the six remaining generating vectors of the modified system are plotted by the curve with 6 points in (c). Further, setting flux #8 in (a) to zero eliminates pathways p_2 and p_5 in (b). The 4 remaining singular values are also shown in (c). Examples (a) and (b) are taken from [Schilling *et al.* (2000)].

unidirectional (positive real number), and less than some unspecified maximum rate. These constraints on flux values require application of a mathematical technique known as *convex analysis* that helps in this case because we know the network topology, viz., Figure 5.8(a). Convex analysis decomposes the network paths into a set of systematically independent *generating vectors*, $p_1, p_2 \ldots p_7$. Generating vectors form a convex basis to represent any input-output relationship that a network in equilibrium can establish. Generating vectors are also called *extreme pathways* [Palsson (2006)], although we prefer the former name as more descriptive.

Generating vectors have become a basic tool for studying systems biology because of the pioneering work of Bernhard Palsson and colleagues. The seven generating vectors for this simple problem are illustrated by the seven paths in Figure 5.8(b). For example, generating vector p_1 connects input entering the system from the left via flux #1 to activate molecule A and then molecule B via flux #2 before exiting the system at the top via flux #3. Since generating vectors span the range of network responses, we say they bound the solution space for modeling. They tell us all possible paths connecting inputs to outputs. These paths are easy to see in small networks, but imagine how important it is to have such an accounting of possible paths in large networks!

The model of Figure 5.8 is too simple to describe realistic networks operating within biological cells. Yet it allows us to peer inside the inner workings of a network that behaves similar to those much larger, and so it is worth a closer look. Palsson and colleagues have successfully scaled up these methods to analyze genome-scale networks of living cells, involving thousands of components [Oberhardt *et al.* (2009)].

Generating vectors are unique and intuitive network descriptors but they are not exactly eigenvectors. Following the methods of Palsson, it is straightforward to apply *singular-value decomposition (SVD) analysis* to the adjacency matrix of a network composed of generating vectors.[5] SVD generates eigen-pathways that are functional network modes describing uncoupled system properties. Generating vectors define possible pathways through the network (Fig. 5.9), like a map of all the roads in town; however, eigenpathways describe their uncoupled functional modes like a map of major thoroughfares through town at rush hour. One describes the possible individual routes through town while the other describes the likelihoods that various combinations of routes will get you to your destination.

SVD analysis provides singular values (SVs); one value for each eigen-pathway. Singular values and eigenpathways serve the roles of eigenvalues and eigenvectors, respectively, for their abilities to describe emergent properties of networks. We computed seven SVs for the seven generating vectors

---

[5]The following might only be helpful to those who know a little linear algebra; our apologies to others. Let P be an $M \times N$ adjacency matrix whose columns are the seven generating vectors $p_1 \ldots p_7$ from Fig. 5.8(b) where all nonzero elements have been set to one. Because matrix P is not square, we pre-multiply by its transpose to find $P^t P$. This matrix product is symmetric, of size $M \times M$, and has rank $r \leq M$ with r nonzero eigenvalues $\lambda_1, \ldots \lambda_r$ that are all real. The r singular values (SVs) of P are square roots of the r nonzero eigenvalues of $P^t P$. We can discuss either the singular values of P or the eigenvalues of $P^t P$ when describing network stability because of their monotonic relationship.

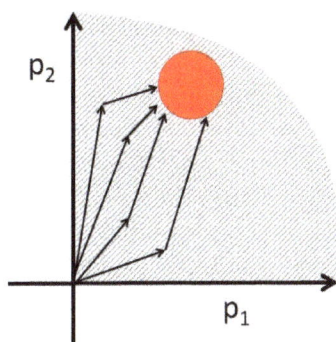

Fig. 5.9 Illustration of a solution space for a small network model composed of just two generating vectors. In contrast, Ex. 1 has 7 generating vectors and Ex. 2 has 325. Four eigenpathways formed in this solution space follow different network paths to give a range of solutions (circle). The shaded quadrant illustrates that the flux positivity has constrained solutions to a region bounded by two generating vectors.

shown in Figure 5.8(b) and plotted them in decreasing order; top curve in Figure 5.8(c). The eigenpathway corresponding to the largest singular value is the principal functional mode of this network.

An eigenpathway is a functional mode of the network formed from a weighted linear combination of generating vectors, as illustrated for a 2-D network in Figure 5.9. Normalized weights on the principal eigenpathway are represented in Figure 5.10(a) by different line widths and numbers listed. Each eigenpathway is a different combination of the same generating vectors, just as each guitar-chord is a different combination of the same vibrating strings. Notice how a "pathway" is really a weighted sum of all generating vectors and not what we normally think of as a single network path. The same point was made in Chapter 3.

The numbers in Figure 5.10(a) describe how well connected each generating vector is to the network; they also describe how accessible a generating vector is for implementing the principal mode. After scaling the principal eigenpathway values so they sum to one, we find that p_2, p_3, p_4 are 2 to 3 times more connected to the network than are the others. Our efforts to minimize mathematics in the discussion discourages further details; please see Schilling's paper [Schilling *et al.* (2000)] or Palsson's book [Palsson (2006)] for mathematical details about generating vectors (extreme pathways).

Now we are ready to purposely modify the network and observe how SVs describe changes in overall system stability. First, we block flux reaction #2 in Figure 5.8(a); i.e., we apply some unspecified drug that prevents that reaction from occurring and thus sets the reaction flux to zero. This action eliminates generating vector p_1 from further participation in network responses, and leaves substrate A disconnected from the network. One of the SVs goes to zero, indicating a pathway has been lost, and the system transitions to a new equilibrium state. We must reapply SVD to the modified network matrix once equilibrium is reestablished to assess the new state of the network. Now there are only six generating vectors and hence six SVs in Figure 5.8(c). All six SVs are slightly smaller than the first six (compare uppermost two curves). The smaller SVs indicate lower network diversity and a less stable network compared to the original.

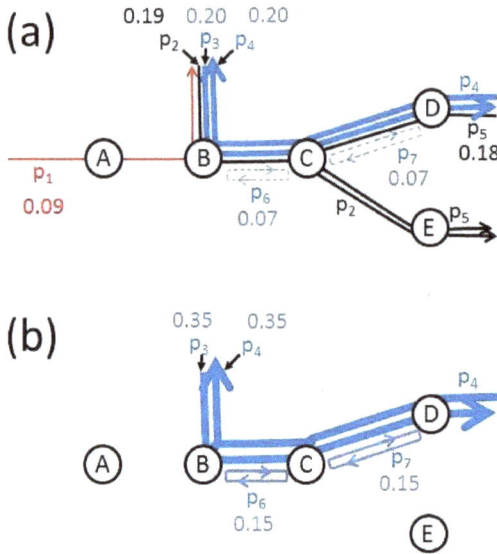

Fig. 5.10   The principal eigenpathway is shown for (a) the network in Fig. 5.8(b) having all seven generating vectors $p_1 \ldots p_7$ and (b) that same network after fluxes #2 and #8 (see Fig. 5.8(a)) were blocked to eliminate p_1, p_2, p_5 so just four generating vectors p_3, p_4, p_6, p_7 remain. Notice how the principal eigenpathways are weighted sums of all generating vectors. The numbers and line weighting given for each generating vector equivalently describe a generating vectors connectedness (accessibility) to the network.

We next disrupt the network further by setting flux reaction #8 in Figure 5.8(a) to zero. This action prevents both p_2 and p_5, the black lines in Figure 5.8(b), from participating in network functions, and thus eliminates any reactions with substrate E. Two more SVs have been reduced to zero by this new action, and again the equilibrium state is changed. The network transitions to a new equilibrium state with the properties indicated by the four remaining nonzero SVs plotted as black dots in Figure 5.8(c). This change further erodes network diversity, substantially reducing the remaining singular values to indicate a less stable network. The accessibility of the network is illustrated with the principal eigenpathway shown in Figure 5.10(b). Compared with Figure 5.10(a), (b) shows the four surviving generating vectors are individually more accessible but the system is overall less stable as indicated by lower singular values. The sum of eigenvalues/SVs is a measure of the resources at the networks disposal to deliver an output. Blocking component connections reduces the available pathways a network has to correctly respond to an input stimulus.

Eigen-analysis of networks is not a new idea. The emphasis on interpretation of eigenvalues/SVs as resource measure is new and needs further investigation. (Cell networks describe highly-nonlinear cell functions, but this graph representation hides the nonlinear elements within the fluxes.) Blocking connections to network nodes can have trivial or far-reaching consequences on network dynamics, and these consequences are difficult to predict except, perhaps, through eigen-analysis. We do not always have a complete mapping of cell network topology as we have in the example of Figure 5.8. Also, we know much less about how cells combine their functions to form the emergent properties of tissues and organisms. Mathematical computation has little power for evaluating eigenstates when our understanding of network structure is limited.

5.10 Implications

The implications of possessing a complete human-system analysis program for medicine are immense. For example, it is desirable in some situations to purposely inhibit cell network responses as a means of curtailing disease progression. Signal inhibition can occur in several ways, including giving patients a drug that prevents ligand binding on specific cell receptors. From the systems perspective, inhibitor drugs lower network eigenvalues for affected cells by reducing the resources that enable a harmful function. Remember that cell networks are the product of millions of years of evolution, so over time they can rewire to restore function and eliminate the blocking effects of an inhibitor drug. This is one type of *drug resistance*. Even if we are successful at turning off specific cell functions, we may cause unanticipated system-wide side effects. The uniqueness of our genomes means that drugs that work well for one group of patients do not work well in many others for a large spectrum of reasons.

As the eigenpathways illustrate, cell networks are rich in *functional redundancies* that allow the emergent properties of cells to remain robust to imposed perturbations. If the patient is fortunate, his/her drug-modified system will turn off specific cell functions without otherwise disturbing homeostasis. However, inhibitor drugs can also prompt cell-network changes that activate new pathways to reestablish the harmful cell function. For example, it will be much more difficult to isolate nodes, as we did for molecules A and E in Figure 5.10, in more realistic and highly-connected cell networks. Therapeutic-drug adaptation is a loss of the intended effects over time as cells adapt to the drug. A drug might also induce changes that kill targeted cells if a vital cell function is lost before the cells can adapt. This is the intention of the marginally-specific blunt instrument known as *chemotherapy* that targets cancer cells hiding in the body after surgical removal of a tumor. Drug adaptations are adaptive properties of cell networks. A model of a patient's networks can help physicians predict drug responses. Eigen-analysis, including SVD, will eventually have a major role in the diagnosis and treatment of diseases involving cellular networks as discussed below.

5.11   Cell Network Analysis: Example 2

Since 1995, Kanehisa Laboratories in Japan have archived cell-network maps and offered them free to the public. From the KEGG Pathways website (see the end of the caption for Fig. 5.11), we acquired a 325-node signaling network segment diagrammed in Figure 5.11. These data provide an opportunity to perform eigen-analysis on a larger cell network segment that was reconstructed from vast amounts of high-throughput genomic sequence data.

In the summer of 2012, the authors worked with a visiting engineering student from École Centrale de Lille, Pierre Pfennig, and with former UIUC Professor Jian Ma and his student Jack Hou to investigate eigen-analysis of this signaling network associated with human cancer cells. The KEGG data describe cancer-cell signaling functions involving a coarse-grained protein-interaction network. We added dynamic features to their static network map by allowing reaction fluxes to vary with time. Our goal was to test the idea that changes in eigenvalues could predict equilibrium-state transitions within a cell.

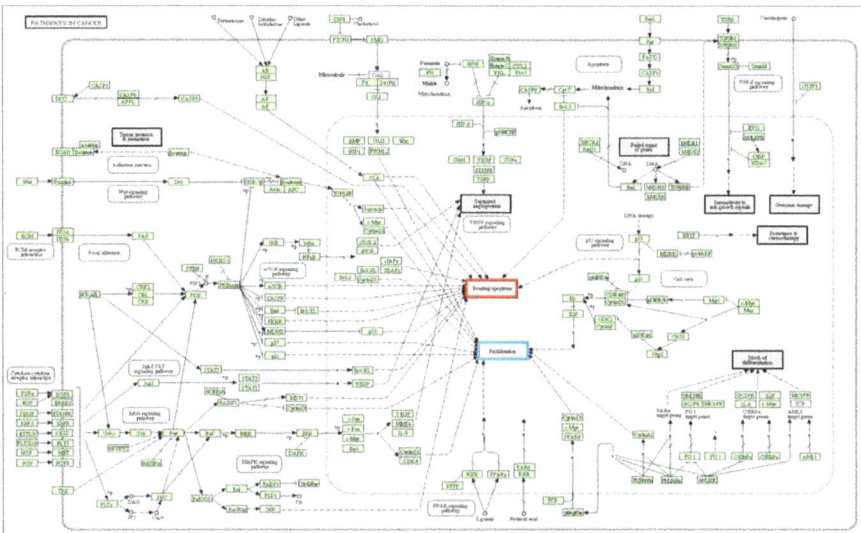

Fig. 5.11 Diagram of molecular interactions within the signaling network of human cancer cells. The diagram was acquired from the public domain KEGG (Kyoto Encyclopedia of Genes and Genomes) pathway database. The cell boundary is indicated by the outer gray line. Green boxes inside the cell are internal reactions and groups of reactions. Input receptors are green boxes within the cell membrane boundary that bind to input ligands (connecting green boxes outside the boundary). The red and blue internal boxes near the center are the output phenotypes that we tracked with our model over time. These are evasion of apoptosis (avoiding programmed cell death) and cell proliferation. Increases in these signaling-network outputs indicate higher likelihood of a cancer-cell phenotype. http://www.genome.jp/kegg/pathway.html.

While repeatedly applying the same random ligand stimuli to modeled cell receptors over thousands of time intervals, we simulated two cellular outputs that indicate the likelihoods of the cell avoiding programmed cell death (i.e., evading apoptosis) and the rate of cell division (proliferation). Among the hallmarks of cancer cells [Hanahan and Weinberg (2011)] is the loss of apoptosis and higher-than-average proliferation, and so these two outputs indicated the likelihood that a cell has a cancerous phenotype. Sudden changes in these two output states with time, either up or down, are equilibrium-state transitions toward a more or less cancerous phenotype, respectively.

In Figure 5.12(a), the number of eigenvalues out of 325 that fell below a threshold value set just above zero is plotted as a function of time. Every 50th time point is plotted, so 10,000 observations are represented by the

200 time points. We did not attempt to accurately model cell dynamics with this model nor did we purposely isolate network nodes as in Example 1; our goal here was to observe a destabilizing transition causes by tiny perturbations in the network topology over time. Between each time step, some of the 325 flux reaction rates of the network were randomly varied a small amount.

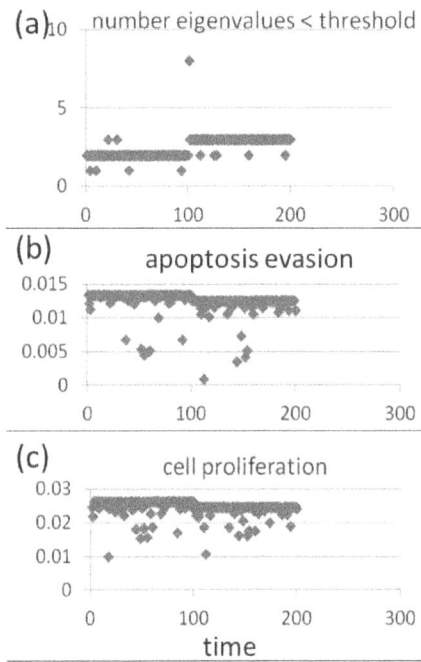

Fig. 5.12 Results of dynamic modeling of the cancer-cell network diagrammed in Fig. 5.11. 10,000 time increments were computed and every 50th point was plotted to give 200 time steps. In (a), the number of network eigenvalues falling below a threshold set just above zero is plotted. In (b) and (c), the corresponding output values are shown. These are output from the red and blue boxes in Fig. 5.11. Near time step 100, the network undergoes a sudden transition to a new equilibrium state.

Initially, two sub-threshold (essentially zero) eigenvalues were found. Initially-zero eigenvalues indicate that not all generating vectors are connected to the network. Until time step 100, the number of sub-threshold eigenvalues varied between 1 and 3, as one might expect given the presence of network fluctuations. Despite local transitions caused when the eigenvalue of a localized eigenvector went to zero, the whole simulated network remained in equilibrium. We know this because both outputs maintained noisy but steady output levels during the first 100 points. As we found with the earthquake example in Chapter 4, transitions can occur locally even though the whole system remains in equilibrium. This tells us the simulated responses are fairly robust to internal variability as one might expect of any evolved networked system. Note that if there was no variability in the fluxes or inputs, the initial output values would not change over time.

Suddenly, after time step 100, eight eigenvalues were found to fall below threshold for just one time step (Fig. 5.12(a)) before settling back to a fairly stable level of three. Simultaneously, we see both network outputs stepping downward (see Fig. 5.12(b) and (c) at time 100). This simulation demonstrates how small topological changes can accumulate over time to destabilize the network until a tipping point is reach and a network-wide transition occurs. A transition has occurred because both outputs suddenly decreased and the average number of zero eigenvalues increased by one, indicating a loss of network connectivity.

We did not plot all 325 eigenvalues over time, although cursory inspections showed a number of them fluctuated as flux variations accumulated over time. At the tipping point, time-step 100, a tiny perturbation plunged five additional eigenvalues toward zero, causing a sudden reduction in the two simulated-cell output levels. The fact that several eigenvalues approached zero for a network-wide transition to occur is interesting. If this transition occurred in a cancer patient, we would interpret the change as "desirable" since properties of the new equilibrium state indicate a "less cancerous" phenotype. That is, after time 100, this simulated cell is somewhat more likely to undergo apoptosis and less likely to proliferate than before the transition. Example 2 frames a new approach to monitoring cancer progression. More fundamentally, this cell-network simulation demonstrates a basic property expected of all complex systems.

We must point out that the transition documented in Figure 5.12 is a rare event. Running this simulation dozens of times, we did not observe a clear transition most of the time, so there is still much to learn about network stability.

5.12 Summary

Examples 1 and 2 above combine to illustrate important properties of perturbed networks. Large state-space dimensionality (many highly-connected network nodes and generating vectors) forms extensive networks that offer vast informational resources for achieving and maintaining a stable equilibrium state. A branch of mathematics known as *combinatorics* teaches us that connections among network nodes grow very rapidly as nodes are added to the network. The potential number of cell functions is staggering given that cells express thousands of proteins! High dimensionality in cellular function demands high energy and raw material costs for sensing, regulating, and maintaining a robust programmed output. Hence many potential cellular functions are deactivated under normal conditions. In resource-rich but variable environments, cellular function can be maintained because of the functional redundancy that can be activated. This capacity provides extraordinary survival advantages to evolved systems.

High dimensionality and variability appear to be important elements of all biological systems, although network variability is a two-edged sword. In our model simulations of Example 2, variability promoted transitions to more desirable states. However, increasing variability a little more sends the model into chaos, where it does not establish an equilibrium state. Were that to occur, a real cell would not survive. Life functions emerge from large, finely-tuned cell networks in ways we are just beginning to understand.

We see the essential value of network diversification in other large systems like economies and ecosystems, and we are beginning to better understand the importance of social and cultural diversity for the existence of a robust human civilization. Our sense is that diversity within all complex systems is essential for stability, which is certainly not obvious at first blush.

5.13   Unique Properties of Living Systems

Before moving into a brief discussion of how systems thinking could influence medicine, we focus on important differences between living and nonliving systems to reflect on important points made by others about modeling emergent properties. Model development is an essential tool for understanding and predicting complex system behavior. However, modeling biological systems requires careful consideration of how their unique properties should influence expectations for and interpretations of modeling results.

Cells in our body, as descendants of replicator molecules, are governed by the laws of physics first and foremost and then by laws of field-tested genomics. At the core of genomic laws is variability in the recording and expression of cellular information. Genomic laws embody our understanding of how the universal features of life that we observe at different scales emerge, from gene expression to organism physiology and even social norms. That's right, some features of social behavior have a genetic basis [Rittschof and Robinson (2016)]. Variability in the transfer of information among cells is an essential part of the life process. It is well known that *chromosomal aberrations* (families of genetic mutations from base-pair deletions, duplications, inversions, and translocations) that drive the evolution of species are essentially molecular copy variations that, over time, became much more than copy errors. Genetic variability generates diversity among individuals of a species that increases the net information contained within their population. Variability also fortifies populations against environmental challenges.

What is less understood is that gene-expression variability provides diversity in ordinary cellular functions, as we have been discussing above. This variability is a good thing because it enables cells to mount a robust cellular response to environmental input variability and changes in network structure. However, this sort of biological variability makes it very difficult to model cell processes using mathematics. Specifically, one input stimulus does not map uniquely to one output response in cell systems. Hence cell functions are not strictly mathematical functions.

Cell inputs entering the network are not forced to follow a specific output solution; they are guided to a range of solutions as illustrated geometrically in Figure 5.9 for a simplified network with just two dimensions. The physical constraints placed on network component operations help define generating vectors that may be linearly combined to predict many different output responses. In this way, we model a range of output responses for each input stimulus. The situation is similar to modeling in the artificial intelligence and operations research fields that try to predict complex system responses using *constraint satisfaction* problem approaches. Valid predictions from constraint satisfaction approaches are those that span the dimensionality of the solution space (e.g., the shaded region in Figure 5.9) while satisfying all functional constraints (represented by the p_1 and p_2 bounds in Fig. 5.8). The multiplicity of valid responses to a single cellular input is what is meant by functional redundancy and cellular multifunctionality. Both are core properties of evolved systems of distributed networks associated with living processes.

The generating vectors of Figure 5.8(b) proposed for modeling cellular metabolism by Schilling, Letscher and Palsson [Schilling *et al.* (2000)] have an important role for building understanding and predictability of cell system behavior. Eigen-analysis of generating vectors explains the relative importance of the available network connections, and it will predict a weakening of the available resources for generating a response that might threaten to suddenly transition system properties into a new equilibrium state. This might form a basis for diagnosis if we know how to read those signals.

Some of you may be saying, we observe variability between input and output in all systems whether or not the system is living. What is so special about living systems? The answer is that variability is built into the cellular machinery and not just a limitation on our knowledge of cellular machinery. Consider standard engineering methods involving measurement instruments. If we know the inputs to the instrument and can measure the corresponding outputs, then we can work from both ends to determine a unique set of instrument properties to within acceptably-small measurement errors when there is a one-to-one relationship between input and output. In contrast, input-output relationships in an evolved network are not one to one because of the adaptive redundancy provided by genomic law.

The best models of cell systems can account for these unique aspects of biological variability. Models based on generating vectors rigorously identify the bounds placed on cellular functions by specifying which molecular mechanisms are possible. Eigen-analysis applied to the generating vectors reveal the uncoupled properties of that cell network. Monitoring eigenvalues as they change with time could provide diagnosticians with rigorous criteria for predicting a weakening equilibrium state in patients. This is true considering that the onset of disease and responses to treatment can be state transitions. A goal of diagnostic systems medicine would be to predict and avoid transitions away from healthy homeostasis. A goal of therapeutic systems medicine might be to induce a transition toward a healthy equilibrium state once an irreversible disease transition has begun. These are engineering methods applied to medical practice. It seems natural for systems engineering to become as fundamental to medicine as it is to architecture, economics and manufacturing.

5.14   Medicine: What Is and What Could Be

Combining all facets of the practice of medicine, we clearly find that the profession of medicine forms a complex system with a very broad-ranging knowledge hierarchy that extends from details of molecular biology to global business strategies and political implications. In the United States in 2016, medicine is struggling to find a sustainable business model whereby society can afford care for the population regardless of an individual's economic status. Throughout the 20th century, advances in technology have been absorbed into the economics of medicine. The result is innovation and improved patient well-being but at rapidly increasing costs. These costs have accumulated to a point where society is now struggling to pay the bill. How can the medical profession continue to generate new knowledge and technology while remaining financially viable and meeting a constantly evolving set of ethical and legal obligations? This is a tough question that will take decades to resolve on our current trajectory.

Many studies have shown there is one change that can improve the stability of the healthcare system: it is to educate people about the consequences of various *lifestyles*. A more knowledgeable public would allow the healthcare system to shift its center from curative to preventative medicine. While it is undoubtedly among the best of ideas for substantially reducing overall healthcare costs and increasing quality of life, successful implementation assumes the population will listen carefully and act accordingly and with reason. Educational processes take time, as we learned from the government campaign against smoking, and there are limits to regulating human behavior in a free society even when the advantages are obvious and when we all share in paying healthcare costs. Most Americans want their medicine at the curative phase and in the form of a safe, effective and inexpensive drug that makes symptoms disappear so we can go on with life and not think too hard about causes and prevention. A large segment of Americans is not fond of assuming responsibility for their personal health when it conflicts with their preferred lifestyle.

It takes as much as a billion dollars and many years for work to bring even one new drug to market largely because the population is risk adverse regarding safe and effective uses of medicine. Despite these costs, some drugs fail to realize their expected effects on population well being until sometimes decades after market release. And why should we expect a single drug to work effectively for everyone given all the biological diversity that makes our species so robust to environmental challenges? The enormous complexity of individual human systems suggests our networks are all tuned differently, and it is only our adaptability that allows one drug to work reasonably well for many. What can system engineering do to improve our approaches to medicine?

In the not-too-distant future, we will be able to sequence each patient's genome for about a hundred dollars. We will also be able to monitor the movement of molecules within and between our cells using an array of new sensing technologies. If we can integrate this information into effective system models, we will have the ability to monitor changes in the human equilibrium state. We believe that a more fully-developed eigen-analysis of the information from properly sampled cellular networks will become the future of medicine. It offers a rigorous engineering framework for analyzing complex systems that could guide financial investments into affordable medical technologies of the future. Since healthcare includes many social, political, environmental and ethical issues that extend well beyond the traditional topics of medicine, the methods of Systems Medicine must expand eventually to encompass these traditionally more peripheral segments.

This general topic was discussed in the May 2014 report to President Obama from the President's Council of Advisors on Science and Technology (PCAST) entitled *Better Health Care and Lower Costs: Accelerating Improvements through Systems Engineering* [PCAST (2014)]. The advisors describe how participant incentives are often at cross purposes with improved efficiency in ways that compromise overall quality and raise healthcare costs. The move to account for all aspects of the healthcare system in its reformulation has begun.

Next, we explore a possible future for Systems Medicine as it might impact one important topic: breast cancer management. The goal is to illustrate how systems thinking that is already common in the science of cancer biology must now expand into medical practice. Cancer is a good example because it is a natural condition arising in most multicellular organisms. It becomes a life-threatening disease when mutated cellular programming couples in dangerous ways to normal cellular information networks to permit a cascade of system transitions to organism-wide scales.

5.15   Cancer

Cancer is a name given to a family of cellular processes that originates when the DNA of a cell in an organism acquires mutant *alleles*. Alleles are alternative forms of a gene that are best known for the observables of dominant and recessive traits. Assume a genetic mutation has turned on expression of normally-quiet proto-oncogenes and/or turned off expression of tumor-suppressor genes. If that cell survives, all of its progeny acquire the genetic programming of a cancerous phenotype. A cancerous phenotype is characterized by "sustaining proliferative signaling, evading growth suppressors, resisting (programmed) cell death, enabling replicative immortality, inducing angiogenesis, and activating invasion and metastasis." [Hanahan and Weinberg (2011)] Healthy cells in our bodies depend on checks and balances that maintain just the right numbers in each organ and prevent cells originating in one organ from growing in another. Cells with a cancerous phenotype are similar to healthy cells except the safeguards are disabled or unbalanced. Unlike a bacterial infection, where the body is invaded by a foreign organism that presents an antigen to provoke an immune response, malignant cancer is a threat from the organism's own cells, which the body frequently does not recognize.

Cancer cells may be characterized by their inappropriate responses to their microenvironment.[6] Furthermore, they can use their amazing repertoire of innate functions to manipulate surrounding cells into promoting tumor growth and metastasis. Tumors are tissue masses composed of cancer cells and support stroma that have a potential to spread through the vasculature throughout the body. Patients die from cancer when their body's burden of supporting primary and metastatic tumors interferes with vital functions of the brain, heart, lungs, liver or other vital organs. Minimizing tumor burden is the immediate goal of cancer treatments. The primary diagnostic goal is to detect a cancerous condition at a stage before it metastasizes, that is, spreads to other organs.

A cell-system description of human cancer progression begins with oncogene activation in one cell. Activation initiates a cell-wide transition from a normal to a cancerous equilibrium phenotype. Many human tumors are initially *monoclonal*, meaning they are copies of that single first cancer cell. However, the process of rapid division of a cell possessing an unstable genome generates further mutations in successive progeny. When this local collection of cells experiences the normal blizzard of signals from a busy epigenetic (non-DNA related) microenvironment, the cancer-cell population diversifies into subtypes with different forms and functions, and these tumors resist treatment.

---

[6]This is a paraphrased statement that Michael Sheetz from Columbia University made at a talk he gave at UIUC in 2012.

We can observe tumor formation by studying cancer cells under a microscope at different stages of development. Successive generations of cells progress through a series of increasingly disordered pathological phases as cells multiply to form a heterogeneous mass with diverse phenotypic sub-populations and high genomic instability. The more heterogeneous a tumor cell population becomes, the more robust its cells are to any immune-system responses and to standard cancer treatments. Once again, we find an example in nature where diversity stabilizes a large complex system. In this case, cellular heterogeneity allows a tumor to remain in a dangerous growth state despite strong environmental influences of anti-cancer drugs intended to disable cancer cells and prevent their spread.

Tumor formation is actually a rare event. Most genetic mutations in cells are repaired by internal cell processes. Those that escape repair to form neoplasms can be eliminated by immune responses without symptoms if the cancer-cell phenotype presents sufficient antigens. Some tumors that escape the immune response can halt their growth process for extended periods of time because of emergent properties of our cell systems that we don't fully understand. It isn't until we reach advanced age or otherwise become immune compromised that genetic errors are allowed to accumulate faster than the error-correction mechanisms and the immune system can mitigate their influence. With safeguards overwhelmed, tumors are more likely to form and grow.

Some patients with a type of skin cancer, melanoma, can respond to treatment and enter a remission phase to become asymptomatic for decades. Then quiescent residual disease hiding in body tissues for many years can suddenly reactivate to become refractory disease (resistant to treatment), placing those patients in extreme peril. What is it about a patient's cell systems that can "turn down the intensity" on the cancerous phenotype sometimes for decades before reactivating incipient disease? The answer is partially that cancer cells require cooperation from natural body mechanisms, like wound healing, if they are to form a tumor able to grow and metastasize. Just as a conquering army needs constant supplies to do its job, so too does a tumor need to subvert body functions for its advancing purposes. Metastasis is the most dangerous stage of cancer, where tumor cells spread to other parts of the body. Maturing tumors can shed cancer cells that enter the vascular or lymphatic circulations to reseed in other parts of the body. About 90% of patients who die from cancer die from metastatic disease.

Metastasis is a strange phenomenon when you consider that it is quite difficult to grow cells from one organ in another organ. Each organ develops its own microenvironment that doesn't nurture cells from other organs, and with good reason. So a circulating cancer cell that finds its way into a different tissue type must coerce its acceptance through the messages it sends via normal cell network communications if that cell is to be implanted and grow. It is common in advanced breast cancer to see metastasis in a patient's bone, lung and brain tissues. One controversial theory is that heterogeneous primary tumors can develop and shed into the vascular circulation *cancer stem cells* (CSC). Stem cells of any type are undifferentiated with respect to their functions and therefore can quickly adopt the characteristics of other cell types. CSCs can quickly adapt to proliferate within different host-tissue microenvironments despite their origin. The cancer biology literature is full of examples where cell signaling of the type illustrated in Figure 1.4 plays a key role in metastatic progression.

5.16   Cancer Progression as Hierarchical Transitions

Cancer mechanisms differ in various organ systems in part because of each system's unique microenvironment. In most focal breast cancers, one malignantly-transformed epithelial cell, often within the lining of a milk duct, initially alters its response to a few environmental signals that influence some of its behavior. The changing cell does not complete its transition to a cancerous phenotype until its apoptotic and proliferative switches are affected.

As illustrated in Figures 5.11 and 5.12 or Example 2 above, when one or more eigenvalues in a cell network approach zero, the response of the cell to the same environmental signals can suddenly reset to new output levels. This is why cancer cells are thought to be responding inappropriately to microenvironmental signals. There are IHC changes[7] in the cell that indicate a phenotypic change has occurred that pathologists recognize when viewing biopsy-tissue samples. Malignant cell transitions can develop quickly or slowly, but they appear to be irreversible, and they can be recognized from their disproportionately-large output responses to small input stimuli.

Over time, as cancer cells proliferate unchecked in a region of tissue, they outgrow their blood supply. Generally, cells must be within about

---

[7]IHC is immune-histochemical analysis of cells extracted from the body (biopsy). It is the process of identifying proteins on and in cells through the use of antibodies that target those molecules as antigens. For microscopy, if the antibodies contain a reporter, then contrast for that protein feature is enhanced for viewing by a pathologist.

0.1 mm of a capillary to access the oxygen available in circulating blood. Low oxygen levels provoke the cells to issue chemical signals that initiate new vascular growth in nearby tissues – *angiogenesis*. The new blood vessels that are formed provide the nutrients needed for additional tumor growth. However, they are poorly constructed and leaky, which offers ample opportunity for CSCs to enter the vascular stream. The growing primary tumor mass and circulating CSCs initiate another organism-wide system transition from the cell to the organ and then organism levels. Most organs in the body provide more functional capacity than is typically needed, so affected organs become symptomatic once tumor damage becomes painful or it reduces organ function below levels needed to sustain health.

Cancerous transitions have been classified into stages that help patients understand their prognosis (disease course). Diagnoses aim to detect disease during the time period between cellular and organ-related malignant transitions, which is the widow of therapeutic opportunity. Stage 0 cancer is in situ disease, which means a neoplasm is early in development at a stage that is not immediately threatening. Stages I, II, and III reflect the increasing size and expanding influence of a tumor. In Stage IV cancer, the likelihood for metastasis is significant. With metastasis, progression begins a whole-body transition into an equilibrium state that poses immediate threats to the life of a patient.

Hence a systems view of cancer is a hierarchy of cell-system transitions to ever larger scales and with expanding somatic effects. The eigenview would say that cancer progression influences ever more-broadly distributed eigenvectors. Indeed, the American Cancer Society quotes the 5-year survival rate for breast patients with Stage II cancer at 93%. The rate falls sharply to 22% for Stage IV patients having evidence of metastasis. Of course, the type of cancer and an individual's genome have much influence on a specific patient's prognosis.

5.17   Diagnostic and Therapeutic Approaches

Until recently, cancer was diagnosed and treated based on the type and stage of disease but without specific information about the individual except family history, life style, and occupational exposures to carcinogens. Today many efforts are being made to use the genetic makeup of individuals that personalize the treatments to be most effective and to pose the least side effects. The challenge is that pre-cancerous in situ disease looks and functions very similar to normal tissues, and therefore it is hard to find and preferentially disable or eliminate. Pathologists routinely "listen" to specific aspects of cellular communication networks to detect and stage the disease by watching which proteins are expressed on cells via expanded IHC analysis applied to biopsy samples. Similar methods are being developed in *molecular imaging techniques* that aim to view protein expression throughout the patient without extracting tissues or fluids. In a system view of these approaches, protein expression is a *biomarker* for monitoring the eigenvalues of the cell system. Although we don't yet know how to estimate a patient's eigenvalues, we do have some of the tools needed to do just that. Identifying biomarkers of diseases like cancer is a major research initiative that promises patient specific diagnostic information about disease states.

Once this diagnostic information is routinely available, patient-specific treatments can be designed and implemented. Discovering appropriate tissue biomarkers for a disease process in a patient is the central problem of systems engineering applied to medicine. Once that information is available, treatments can be designed to "tune" the tissue microenvironment so that the natural immune system and other micro-environmental factors can eliminate or hold in check cell-network diseases like cancer. This is a very exciting approach because, if successful, it promises to make cancer a controllable condition rather than a life-threatening disease. It also seems that systems engineering methods applied to medicine would be more economical if we can more fully exploit our ability to appropriately sample the human system using advanced molecular diagnostic technologies under the guidance of each patient's genomic signature. Considering the high cost of targeted diagnostic and therapeutic agents today, systems engineering methods can provide a more directed approach to drug development, selection, and timing for maximal therapeutic effect and minimal side effects.

It is true that each cell's activities are partially regulated through precise cooperation of its networks with the trillions of other cells in the body that is strongly influenced by individual lifestyle and living environment. It is also true that the sheer magnitude of the network assessment task is what has largely thwarted the "War on Cancer," which was a sincere attempt by government agencies under the Nixon administration to pass the National Cancer Act of 1971. It provided national-scale resources to help medicine understand the nature of the disease and implement those discoveries into clinical practice to reduce cancer deaths. What we didn't know then was that cancer is not "a disease" like other diseases and therefore does not have "a cure" like other cures. Cancer is a large family of multifaceted conditions with many causes and progression mechanisms. We were hoping to discover how Santa Claus was able to deliver all the toys on Christmas night and instead discovered a highly-distributed global retail system of manufacturing, shipping, financing, etc. with complexity beyond imagination.

In 2016, a consortium of industrial (including pharma), medical and academic organizations has proposed the "Cancer MoonShot 2020" program. Hoping to leverage advances in precision medicine where a patient's immune system is activated to treat cancer, the goal is "to accelerate the potential of combination immunotherapy as the next generation standard of care in cancer patients." Even if these efforts do not result in cures, it is now reasonable to adopt the ambitious goal of at least managing cancer as a survivable condition. This view recognizes that some cancerous phenotypes may be with us periodically throughout our lives, and need to be managed systemically as we do for many with atherosclerosis and diabetes. We need to define coarse-scale biomarkers that indicate changes in human-system eigenstates specific to disease conditions. This approach might not work for all conditions, but it certainly looks promising for treating cancer and any other disease of cellular networks.

As compelling as we find the *Systems Medicine* approach for the future of medicine, we also are aware of many formidable challenges. We note just a few.

5.18   Conflicting Signals

We now give a well-known example of just how complicated the system of cells in our bodies can be when it comes to cancer. One of the oncogene activation processes involves the response of mammary cells to members of the *transforming growth factor beta* (TGFβ) family of signaling proteins. Specifically, TGFβ-1 controls normal breast morphology, including the suppression of neoplastic growths. It regulates the cell cycle and induces apoptosis when needed. In a normal equilibrium state, TGFβ-1 also appropriately minimizes mammary-cell proliferation. However, when mammary cells with switched-on oncogenes are exposed to the same levels of TGFβ-1, the response is now aggressively proliferative. The malignant transition of a mammary epithelial cell diametrically changes the effects of TGFβ-1 on the cell from a potent tumor-cell *suppressor* in normal cells to a potent tumorigenic-metastatic *promoter* in cancer cells. This effect is from cell network changes induced by the malignant transformation. We need to understand this type of multifunctionality at work within cellular network functions. What is nature thinking!

5.19 Treatment Resistance

Standard breast-cancer treatments include combinations of surgery, chemotherapy and radiation therapy. Each aims to preferentially kill, remove, or deactivate malignantly-transformed cells while creating as little damage to noncancerous cells as possible. Since cancerous and noncancerous cells have a similar phenotype, the therapeutic margin is slim under the best conditions. In heterogeneous tumors, chemotherapy can initially reduce tumor mass by eliminating some subtypes until the only cells that remain are resistant to treatment and continue to proliferate. The diversity of cell subtypes in many tumors prompted oncologists to minimize single-approach treatments in favor of those that diversify the therapeutic attack.

Adjuvant therapies combine treatment options to eliminate cell types not affected by one option. For example, studies showed that for Stage I-III breast cancers, surgical lumpectomy followed by combinations of photon radiation, hormone therapy (e.g., tamoxifen), signal inhibitors (e.g., the monoclonal antibody trastuzumab, better known by the brand name Herceptin), and chemotherapy (e.g., doxorubicin, paclitaxel) was equally or more effective than total mastectomy for many patients. Some cancer treatments also involve the use of monoclonal antibodies during a process that labels cancer cells as foreign so they will be targeted for elimination by the patient's immune system.

The high cost of medicine tends to promote techniques that treat the "average patient" rather than individuals. This approach is less effective at treating each specific patient because of a broad diversity that exists in patient physiomes. Finding one treatment that works routinely and predictably for all patients is rare. The hope for the future is to be able to read a patient's genome and to combine that information with details of their home and work environments, family history, and lifestyle using data-driven modeling to learn which prognostic and treatment responses are likely under different scenarios. If successful, personalized medicine techniques could quickly identify the most effective treatment for each person and at lower cost to the healthcare system and therefore all of society.

5.20   Eigen-analysis in Medicine

In earlier chapters, we showed that as eigenvalues approach zero, the corresponding system can undergo a transition that establishes a new equilibrium state with potentially very different system properties. Some system transitions may only influence local properties significantly even as the whole system undergoes a transition. However, if the whole system has small eigenvalues indicating instability in its equilibrium state, then any small perturbation can initiate a transition that propagates in scale to greatly influence all aspects of the larger system. The propagation of a transition to larger scales within a system provides analysts with some time to diagnose its occurrence and possibly design techniques to effectively intervene, as we discovered when exploring the systems view of cancer progression.

We can look for opportunities to apply systems analysis in medicine by noticing analogies found when observing other complex systems. For example, it seems to us there could be parallels between the behaviors of cell networks and the Tacoma Narrows suspension bridge. In the case of the bridge, when wind blew across the deck, stability was lost although no eigenvalues went to zero. As some eigenvalues became equal to each other, forming a subspace, we saw that forces applied to one of the bridges eigenmodes could transfer to another in the subspace, which ultimately destroyed the bridge. Subspace formation seems possible in cell networks, which could help explain how functional modes in cells link together and disassociate to generate unexpected nonlinear responses.

We believe there are general properties that apply to all complex systems, and that once we establish what they might be a gold mine of solutions will open up to us. Hence research conducted to characterize any one complex system will have important implications to all others; we just need to be ready to find and interpret commonalities using scientific analogy. The label of systems engineering has been with us many years, but only now is it clear how important and far reaching the science of systems engineering can be to our future; especially in the study of life – the most complex of all systems.

Chapter 6

Role of Variability in Systems

6.1 Introduction

For a large collection of component nodes to behave collectively as a complex system, each component must be in communication with many of the others. The communication channels that connect components can be clearly physical channels (involving straightforward mass-energy exchanges) or less-clearly physical channels like those found in financial markets or emotional exchanges made between people. When component connectivity is high, individual component responses have an opportunity to combine and generate *coherent responses* that are characteristic of the emergent properties of a system. This chapter probes the essential role of component-property variability in creating the characteristically different properties that emerge in *homogeneous* (simple) systems and *heterogeneous* (complex) systems.

Simple systems often have centrally-connected and controlled components. This occurs in a network by restricting the flow of information to components through an external central-processing hub (Fig. 1.6). Chapter 1 gave an example of a simple system using the rail-transportation network. There might be many trains (components) in the railway system, but their central control restricts the ways those trains can move so that that the system has few degrees of freedom. Centrally-controlled networks have important advantages over self-organized networks when predictability of system behavior is valued over adaptability to a changing environment.

Broadly-distributed connectivity is necessary but not sufficient for complex-system behavior. Properties of complex systems are created by a "large number" of *degrees of freedom* within the component network. One major goal of Chapter 6 is to explain this statement at several levels.

Degrees of freedom determine the number of different responses that a system can generate after receiving a stimulus. The self-organization and high connectivity of complex systems make them capable of broad ranging responses. Nature is full of complexity, as described in Chapter 1. It is the models we build to study these phenomena that do not always behave as complex systems. We often choose to model parts of natural systems by abstracting those features we wish to study and ignoring the others. These models offer valuable insights about specific mechanisms at work in the system but they are limited in their ability to generate the full range of responses that are seen in natural systems.

To illustrate, consider an intuitive dynamical model of cyclic population growth and decline from the field of Mathematical Biology – the classic *predator-prey model*. The basic form of this mathematical model describes temporal changes in the numbers of two mutually-dependent animal populations collocated in one region. One species preys on the other in a manner that induces cyclic changes to the coupled population numbers about equilibrium values. We will discuss fox and rabbit populations, but extensions to viruses in human populations are not difficult to imagine. The classic formulation of the predator-prey system has members of each population exhibiting *homogeneous properties* that are defined in the next section.

These population models are taught to students as a demonstration of emergent behavior of co-dependent identically-distributed populations. You see, the two equations each with two variables form 2×2 matrices that have easy-to-calculate and easy-to-interpret eigenstates! Yet the equations do not represent a natural complex system. Regardless of the population sizes, as long as the model has only a few degrees of freedom, its predictions lack the range of responses found in nature. More representative models will include subpopulations with distinct properties (represented by more equations and variables) that increase the number of degrees of freedom in the system. Subpopulations gives rise to the emergent properties expected of a dynamic equilibrium state. But what sorts of variability are required for this to happen?

6.2   Degrees of Freedom, Accessibility, and Emergence

Homogeneous populations and parametric variability.   A modeled population of foxes interacts with a modeled population of rabbits as diagrammed in Figure 6.1. At this point, assume each population is individually *homogeneous* in the sense that properties of one fox are identical to all others. Similarly for all rabbits, and each fox-rabbit interaction is the same for any property that influences the model equations. For our discussion, assume that groups of exactly eight foxes band together to capture and eat exactly one rabbit over some time interval. Which rabbit? It doesn't matter because they are all the same. During the same time interval, exactly one rabbit is born for each one that is eaten. At the moment a rabbit is born, it is instantly an adult with the same properties as all other rabbits. Individuals change but population properties remain the same day to day.

Not very realistic, huh? We agree. This model is a rather extreme *abstraction* of a real ecosystem that has great value for teaching and achieving basic understanding of population dynamics. We say such models are mathematically *tractable* and *transparent*, and we accept that the model sacrifices accurate predictability for some details in order to offer broad intuition about system behavior.

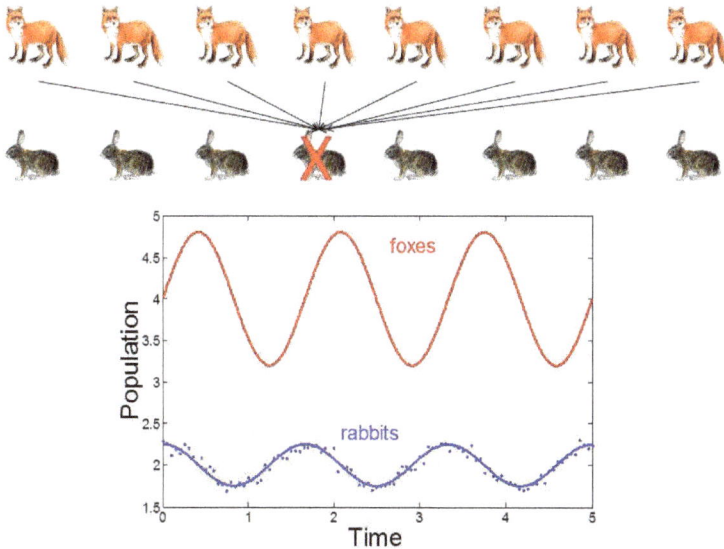

Fig. 6.1 (Top) Illustration of changes in two homogeneous populations that interact over time. In this model, groups of foxes capture and eat rabbits. (Bottom) If we let the rabbit birth rate and species-interaction rate differ, and we include resource limitations and natural death rates, the two populations are found to oscillate in time at the same frequency but with different phases. If we add probabilistic variability to a small rabbit population we will measure the dots surrounding the rabbit-population line, but if the population is large enough the solution will converge to the smooth line. The situation is similar for foxes.

The homogeneous system of two species is represented by two equations,[1] one for each population variable. There are also constant parameters describing the rabbit birth rate, the fox-rabbit interaction rate, and other effects like natural death rates and resource limitations. If the rates at which rabbits are eaten differ from rabbit births, the population time series, shown by the two smooth lines in the lower panel of Figure 6.1, will oscillate about different equilibrium values. The equilibrium populations shown are 4 units for foxes and 2 units for rabbits.

Following changes in *model variables* is the point of modeling, the thing we hope to predict. Variables depend on model *parameters* and *time*. Setting the parameters to be constants yields the simple linear model we want. However, parameters will become functions of each other and time in more realistic, nonlinear models. Let's maintain model linearity for now but we will add parametric "variability" to see what happens.

If constant parameters are replaced by random parameters, then *on average* one rabbit is eaten for every eight foxes in the community per time interval. We might intuitively consider random parameters to be more representative of natural systems, but this type of variability does not fundamentally change the nature of the model; there are still only two population variables. In Figure 6.1, temporal changes in the "measured" population for a relatively-small rabbit community are shown by the dots around the curve, while the normalized population for the larger rabbit community is shown by the smooth curved line. The continuity of predictions over time depends on details of the probability functions adopted.

This linear model describes a homogenous system where properties are everywhere the same. It predicts simple emergent properties,[2] like temporal

---

[1] For mathematically-inclined readers, the two equations are coupled first-order differential equations with constant coefficients of the form $\dot{N} = a_1 N - a_2 N^2 - a_3 MN$ and $\dot{M} = a_4 MN - a_5 M$, where $a_n \geq 0$. $N(t)$ and $M(t)$ are time-dependent population variables while $\dot{N}(t)$ and $\dot{M}(t)$ are temporal derivatives of those variables. Combined into vectors, $\mathbf{N} = (M\ N)^t$ and $\mathbf{a} = (a_1\ a_2 \ldots)^t$, the equations are expressed compactly via nonlinear vector functions $\mathbf{f}$ via, $\dot{\mathbf{N}} = \mathbf{f}(\mathbf{a}, \mathbf{N})$ having solutions $\mathbf{N}(t, \mathbf{a}) = (N(t, \mathbf{a})\ M(t, \mathbf{a}))^t$. If $a_2 = 0$ we have the classic *Lotka-Volterra system*, and if $M = 0\ \forall t$ we find a classic *logistic equation* solution for the single population $N(t)$. This quasi-static nonlinear "system of equations" with two degrees of freedom describes populations that do not form any complex system we would find in nature. Still, it is a great first step in mathematical modeling of populations.

[2] The simplified emergent properties of the classic predator-prey model are summarized by the frequency, amplitude and phases of the populations. The two distinct eigenvalues for this 2-D system of equations are determined by parameters $\mathbf{a} = (a_1\ a_2 \ldots)^t$ that set the frequency and relative phase of the population oscillations. Their eigenfunctions are sinusoids.

oscillations in population numbers and relative population stability similar to that seen in nature. However the model cannot predict sudden extinctions or other major equilibrium-state transitions. For this to occur, the model must allow system properties to evolve over time. Something fundamental is missing.

To find out what, we look closely at what occurs when different types of variability are introduced into these networked components. The random parameters used above and in Figure 6.1 are really just a coping mechanism for managing our lack of detailed knowledge regarding properties of individual animals and their environment. Random parameters increase the number of degrees of freedom a small amount because the output can now assume more states. The missing information about subpopulations is knowable but at the time of modeling it is unknown for any number of reasons. We might assume terrain resources are constant when in reality there are better hiding places in some regions that we haven't taken time to measure. Also some rabbits or foxes are likely to be faster or more clever than others. Modeler ignorance of ecosystem details may be included somewhat by adding random parameters, allowing the population to fluctuate similar to that observed experimentally. These details are unfortunately superficial.

A natural ecosystem has at least as many eigenstates as there are animals, and probably many times more. However, the homogeneity of the animals provides only two *accessible eigenstates*[3] The model is flawed because its *dimensionality* does not match that of the natural ecosystem.

By glossing over population details, we ignore essential system information provided by the additional dimensions contributed by each subpopulation. If there are more than two variables, the model may miss predicting the emergent properties generated by the coherent interactions taking place among subgroups. Let's examine the eigenstates of this two-variable homogeneous model before examining methods for including subpopulations.

There are at least as many eigenstates as there are components in this system, but only two states are accessible in our simple population model to provide output responses to input stimuli. We can find the two nonzero eigenvalues by computing linearized forms of the nonlinear equations at each point in time near the equilibrium populations. The eigenvalues depend only on the rates of rabbit birth and fox death. (You need to calculate them to see these parametric influences.) Given that properties of animal

---

[3]There are only two nonzero eigenvalues, one for each model variable, and two linearly independent eigenfunctions.

behavior are modeled mathematically using equation parameters, it makes sense that the eigenvalues are composed of these parameters. The many other eigenstates, those with zero (or near-zero) eigenvalues, are inactive – inaccessible.

We selected parameters so the populations oscillate about specific equilibrium values.[4] The rabbit population in Figure 6.1 is not resource limited (the model assumes rabbits have plenty of food and there are safe places to live for all) and model parameters do not change over time. Consequently, as rabbit numbers increase, a short time later fox births increase with their growing food supply. The surging fox population causes the rabbit population to fall, which later causes a decline in fox population from their food becoming scarce. A reduced fox population leads to a resurgent rabbit population, and the cycle repeats. These are the time-series results of a linearized system of equations. The more general nonlinear model predicts similar oscillations provided the populations do not stray too far from equilibrium. We can compute eigenstates for linearized versions of the model.

One way to do this is to periodically stop the model in time and instantaneously linearize the equations. We teach students how to do this with Taylor series expansions. Linear models are great for teaching, nonlinear models are more accurate for prediction, but neither describe natural systems until the number of degrees of freedom in the model is close to that found in nature.

Using the linear model, we predicted the responses of the system shown in Figure 6.2 to a specific type of input stimulus. The stimulus was a sudden reduction in the fox birth-rate parameter at time 1000 that modifies the eigenstates (system properties). This is indicated by the changes in oscillation frequency, phase and amplitude for both populations. The oscillation amplitudes increase, and, as expected, the equilibrium population for foxes is reduced but not that for rabbits. Thereafter the eigenvalues are again constant in space and time as the populations settles into new equilibrium

---

[4]The two accessible eigenvalues for the system of Fig. 6.1 are imaginary numbers. That is, the real parts of the eigenvalues are both zero and this homogeneous system is unstable. The simplicity of the system means that if parameters changed so that the real parts of the eigenvalues became slightly positive, both populations spiral up and out of control; slightly negative and the populations spiral to zero. That does not happen in nature, just in our simple model of nature. Here population changes are entirely predictable and proportional to changes in parameters (Fig. 6.2). Since the real parts of the coupled eigenvalues are zero, the system is at a stability threshold. Yet there is no sudden transition because the responses remain proportional to the input stimuli.

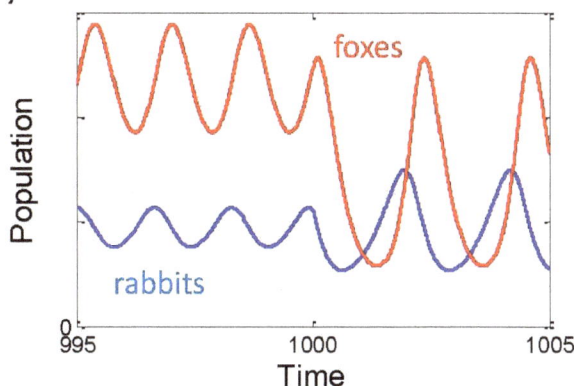

Fig. 6.2   Time series for predator (fox) and prey (rabbit) populations are shown. At time point 1000, we modified model parameters for the fox population only, which modified the frequency and amplitudes of both. This homogeneous system responds predictably according to the properties of any one fox-rabbit pair.

states. The response of the system to varying parameters is not a transition despite there being a change in equilibrium states because the changes are in proportion to the stimulus. Since model predictions for a homogeneous system of equations apply everywhere in the population, at all spatial scales equally, the model is said to be *scale invariant*. That is, predictions made for any small region apply equally to the whole population.

Variability that leads to heterogeneity.   What can we expect to change in the way we model the populations if, say, half of the rabbits and half of the foxes are much faster than the other halves? Also let's assume that resources in the terrain have spatial features that influence the populations. We still have two species, but now there are four populations each with distinct and known (or assumed) properties that can take advantage of the varying environmental resources in different ways. The expanded model now has twice the number of degrees of freedom quantified by and equal to the number of nonzero eigenvalues. Since degrees of freedom quantify the ways a system can respond to stimuli, they also describe the number of factors influencing the establishment of an equilibrium state. When system eigenvalues are distinct, each is a unique model-parameter combination determining the stability of the system's response to a intrinsic mode of activity (system property) that we find from the associated eigenfunction. In our predator-prey model, system subpopulations are model variables that define the equilibrium state of that system at each moment in time.

Equation parameters may be codependent as faster animals are more likely to remain alive and reproduce than slower animals. Small variations in component properties mean these subpopulations are more likely to be spatially variable across the territory. Where we find a high density of fast animals, the eigenvectors associated with distinct eigenvalues concentrate their influence regionally. The localized nature of the two now-accessible eigenstates means the system is losing scale invariance with respect to system properties. An eigenvector describes how an uncoupled system property is distributed across system components. These changes make the model a little more realistic and a little less tractable, given that there are now four accessible eigenstates that influence behavior.

We have now added *component-property variability* to the model based on some observations we made about natural systems. This is better than adding random parametric variability that really only reflects our ignorance about the natural system we are trying to model. Component-property variability increased the number of model variables reflecting more model dimensions. Insofar as the observations are accurate, we have taken a fundamental step forward in developing a more representative model. The natural fox-rabbit ecosystem has this variability, there is no doubt; the only choice for us is how much of this ecosystem variability can and should be included in the model to achieve the modeling goals. Note that adding many types of known variability to the model makes it more difficult to understand the underlying mechanisms of population dynamics. The reductionist in us recoils in worry about an unwieldy mathematical model. Our decision to add known component-property variability influences both the accuracy and the transparency of model predictions when scaled up to a realistic population.

Variability and diversity. We quickly divert the reader's attention to a discussion of differences between variability and diversity. *Variability* refers to differences among properties for a given component type while *diversity* refers to there being different types of components with different component properties. Variance in the predator-prey model can occur from differences in individual component skills, ages, and wellness among a type. However, we might also diversify the predator types by including humans, coyotes and infectious bacteria along with the foxes, each contributing distinct properties according to the species. Things get interesting if we realistically account for inter-predator interactions. An individual might be a predator in one interaction and prey in another. It seems to us that diversity and variability are distinctions with little difference in regard to complex-system formation. Both can add to the number of degrees of freedom and both enrich the number of scale-varying emergent properties in the model. Whatever they are called, diversity and variability in component properties, which are fundamental to all natural systems, must be considered in building accurate models.

Highly-heterogeneous populations. Returning to the population example, imagine a model that realistically accounts for the unique component properties of all population members as might be expected given natural genetic variability driven higher by variability in environmental resource inputs. It is not hard to accept that realistic systems contain a broad range of variability. What is unusual about this particular modeling situation is that we somehow know all of these many properties well enough to accurately model them. These rabbits and foxes would develop a broad spectrum of individual speeds, hiding abilities, successes in food gathering, etc., as well as much variability in proliferation, interaction and survival rates. Diversification results in regional concentrations of animals with similar skills. Each subpopulation seeks an environmental location that best leverages their common skillset, where the goal is to gain a survival advantage.

For example, those rabbits best able to hide might concentrate their numbers in moderately resourceful terrain with low predator exposure, while others skilled at escaping capture can venture into more resource-rich terrain where predator exposure may be greater but the challenge is manageable. Niche communities naturally self organize based on common or complementary properties among individuals. There is no central control telling foxes or rabbits where to live or eat; subpopulation properties emerge naturally and at various spatial scales.

Combinatorics dictate that a great many equilibrium states are possible from a highly-variable community of animals each with distinct properties yielding at least as many degrees of freedom and distinct eigenstates. Most or all of the eigenstates are accessible in the sense that they express system properties. It now matters greatly which individuals are interacting at every location and at each point in time if our goal is to predict population numbers. Component-property variability has increased the number of variables as well as the number of factors determining the equilibrium states this heterogenous population can assume. Variability increases system *entropy* as it adds the degrees of freedom necessary to express the full range of emergent properties of a natural complex system.

To model these populations mathematically[5] we must include an equation for every accessible eigenstates if we hope to accurately predict the equilibrium states of animals in the population.

What is different between modeled heterogeneous and homogeneous populations is the extent of our knowledge about them and the range of emergent properties expressed by each model. Nature is replete with complex systems having many degrees of freedom. Modelers must determine the minimum dimension required to address the modeling question, realizing that highly homogeneous systems exist only in the simplified world of modeling. We cannot be fooled by the lure of intuition offered by these simple mathematical models that let us peer clearly into the inner workings of simple systems; nature is much messier, opaque, and way more interesting. Homogeneous-system models provide qualitative intuition but lack the range of emergent properties expressed by natural systems.

Properties of self-organized subpopulations emerge from models if component-property variability is accurately represented. However, the spatial influence of each subpopulation can vary widely. For example, there might be eigenvalues primarily associated with well-hidden rabbits that cluster tightly in a thicket. Population modes associated with these eigenvalues are described by eigenvectors having local influence, which we label *local eigenvectors*.

Different subpopulation properties may have broad spatial influence. Eigenvalues expressing global properties are associated with *global eigenvectors*. The two equations describing strictly homogeneous populations describe overall properties, which is why we refer to them as scale invariant. An important property of systems with highly-variable component and interaction properties is that *scale-varying emergent properties* form. Each eigenvalue extends its influence on network properties according to the spatial extent of its corresponding eigenvector. A spatial hierarchy of eigenvector influences is a characteristic feature distinguishing heterogeneous and

---

[5] Again, for the mathematically-oriented reader, variability in mathematical models can appear within parameter vector $\mathbf{a}$ whose elements depend on spatial location x and time t as well as population variables $\mathbf{N}$. Changes in heterogeneous populations may be described by fully nonlinear ODEs, $\dot{\mathbf{N}} = \mathbf{f}(t, \mathbf{x}, \mathbf{a}, \mathbf{N})$. Solutions have the form $\mathbf{N}(t, \mathbf{x}, \mathbf{a})$. Near equilibrium, phase portraits for the nth subpopulation can be viewed as plots of N_n versus M_n. The corresponding time series are $N_n(t)$ and $M_n(t)$ as shown for homogeneous populations in Fig. 6.1

homogenous systems. We will see later that scale-varying eigenvectors add stability to complex-system equilibria.

6.3 Local and Global Eigenvectors

We showed in Chapters 4 and 5 that eigenvectors connect every component in a system more or less. The influence of an eigenvector in a region of the system domain is a matter of degree. For networked components, eigenvector influence depends on the patterns of connection weights that define modal properties. Thinking back to the guitar-string example, all parts of the string move with every mode of vibration that generates sound. Depending on the mode, however, some sections of the string move more than others, which indicate how eigenvector weightings link components. Component properties of the strings, guitar body, player and music venue determine the eigenvalues of that system, while the degree to which a component is involved in generating uncoupled properties depend on the vibrational mode – the eigenvector. Eigenvectors are modified, just as eigenvalues change when properties of nonlinear systems change, either through slow evolutionary pressures or suddenly following a transition.

In the population example, assume a clever group of aggressive foxes discovers a sizable community of well-hidden rabbits and wipes them all out. That event prompts one or more eigenvalues of the whole ecosystem to fall toward zero, inducing a system-wide transition to new equilibrium states. The effects of this particular transition might be so localized as to have minimal effect on the whole rabbit population, or the transition could be catastrophic if for some reason the affected eigenvectors are global.

In this way, high variability in component and interaction properties can either attenuate or amplify the overall system response to a state transition event. Large complex systems with many more local eigenvectors than global eigenvectors[6] are overall more robust to transitions than small complex systems. We believe this observation was on Ashby's mind when he formulated the "law of requisite variety" nearly 60 years ago [Ashby (1958)].

The consequences of the scale dependence of eigenvectors were illustrated in Chapter 4 in a discussion of earthquakes. During a Napa Valley quake, eigenvalues for the whole-Earth system approached zero, but the associated eigenvectors were localized to the North-Central California region and so the damage remained localized. Most of the Earth was unaffected

---

[6]Self-organized critical systems discussed at the end of the chapter are one example.

even though that small tectonic interface is mechanically connected to every part of the planet. If we had a mapping of all eigenvectors in a system, we could predict the effects of a transition by studying the extent of the affected eigenvectors. In contrast, eliminating some percentage of components in a homogeneous population can only have a proportional effect on the population as a whole. Homogeneous systems are way more predictable than complex systems.

6.4   Summary

A complex system has many components that are widely interconnected but not centrally controlled. It must also have many degrees of freedom that indicate the number of accessible eigenstates responsible for generating a spatial hierarchy of coherent system properties. Natural systems often demonstrate much variability among the individual-component properties and their interactions. This type of variability gives rise to high-dimensional complex systems with scale-dependent emergent properties. Scale-dependence is a consequence of uncoupled emergent properties associated with eigenvectors having a spatial hierarchy, some highly localized, others with global influence, and still others somewhere in between. These are characterizing features of natural complex systems.

Simple mathematical models describe abstractions of complex systems that do not exist in nature. They are useful for describing the mechanisms at work within a natural complex system, and they can predict some qualitative features of the responses to input. However homogeneous-system models generally lack full information about exact network topology, detailed component features, or environmental influences. Natural systems are complex because component variability is the norm. Injecting probabilistic variability into low-dimensional mathematical models may predict system responses similar to those found in nature, but they will never predict the full range of emergent-property responses and therefore are fundamentally limited. There are fundamentally different types of variability. Are there modeling options? Absolutely! It is possible to build computational system models in much the same way as they form in nature.

6.5   Mathematical and Agent-based Models

With maximum variability in the predator-prey model, where there are N prey and M predators each with distinct properties, there will be at least

$N + M$ state variables and distinct eigenstates. The variables are no longer the numbers of individuals; now we select metrics related to individual health, like cardiac output or potential for producing progeny. To model this situation, we need a system of $N + M$ state equations with on the order of $(N + M)^2$ parameters.[7] Each state variable changes with time. To mathematically model a system where N and M are large, we will have our hands full measuring and testing all the equations and their parameters. And even if we are successful at building a model with each detail correctly identified, a nonlinear system found in nature will have time-varying parameters that evolve. We could cluster individuals into subgroups for coarse-grain modeling, but that approach has its own strengths and weaknesses. Complex adaptive systems are fast moving targets for modelers.

An alternative, frequently used approach is agent-based modeling first described by John Holland [Holland (1995)]. These models include an agent for each component (components may have many degrees of freedom), and each is governed by rules found empirically by observing natural systems. As with dynamical equations, where changes in state variables are computed at each time increment, agents execute a unique set of rules at each time increment according to the input signals it receives from other agents and the environment. Agent-based models don't offer the same degree of intuitive transparency as mathematical models, but agents can evolve in response to the environment and they form subgroups very much as we see in nature. Observations of agent-based models reveal ensemble behavior just as we find in classical biological and social science experimental investigations.

Agent rules are found using the simple patterns of biological/social behavior observed in natural systems (Fig. 6.3). An example was given in the description of flocking-bird models in Section 1.2. If there are many agents, each with memory, a unique and adaptable rule set, and an opportunity to freely interact with other agents and the environment, we will see system-level properties emerge. Properties determine the system responses to stimuli. Applying agent-based models, we avoid the need to first know details of each component mechanism. Instead, interactions among agents that occur during the modeling process (Fig. 6.3) will reveal underlying mechanisms of the system if we are watching closely.

We know how to compute eigenstates of mathematical models, which is why they are preferred when they can be accurately formulated. We

---

[7]These parameters are entries in a Jacobian matrix that map input factors like available food and weather conditions into output variables like health metrics.

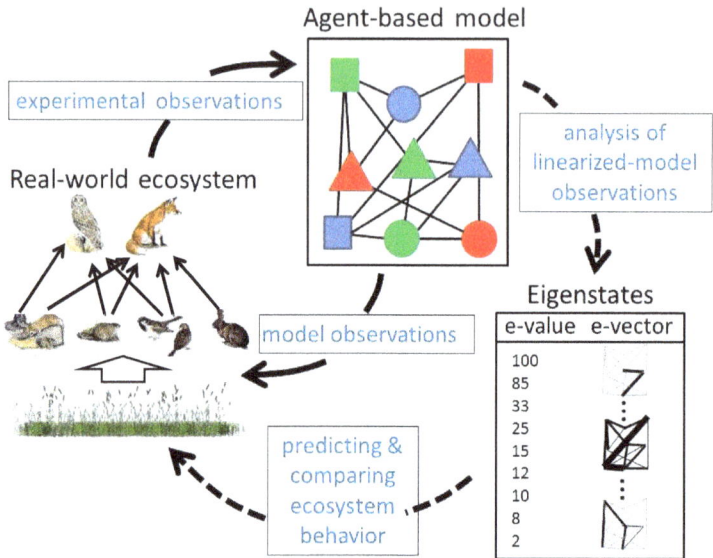

Fig. 6.3   Iterative development of an agent-based model (solid arrows) and eigenanalysis (dashed arrows) of a fox-rabbit ecosystem are diagrammed.

don't yet know how to compute the eigenstates of agent-based and other computational models, but experience gives us enough confidence to conjecture it is possible. Research into agent-based models is an active area for many in industrial engineering and social science fields that hope to find principles and procedures for managing very large complex adaptive systems. Agent-based approaches are now being investigated to model many networked systems in virtually all fields of study.

6.6   Equilibrium-State Transitions Revisited

Homogeneous system example.   All systems can change their equilibrium state and therefore all systems can undergo a state transition, but transitions in simple homogeneous systems can be very different from those in complex heterogeneous systems. Among the differences, transitions in homogeneous systems are more predictable and often globally transformative than those in complex systems because highly-connected homogeneous systems have only a few *accessible eigenstates*. When system eigenvalues approach zero, the observable effects on a functionally low-dimensional system are often globally dramatic and mathematically predictable.

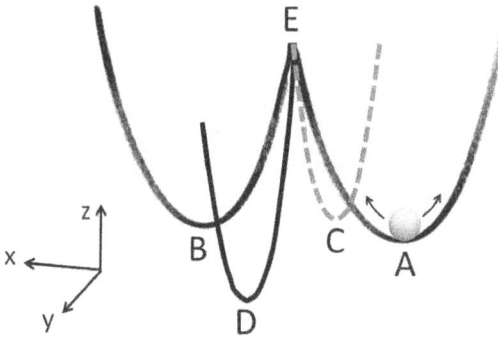

Fig. 6.4   A homogeneous mechanical system consisting of a ball currently at A that can roll on one of four track branches.

To illustrate simply, consider the mechanical model of Figure 6.4. A ball at position A that is free to roll in the x, z plane along track branches A and B or in the y, z plane along branches C and D. To move between track branches, the ball must acquire the energy to pass point E while staying on the track at all times. In our imaginations, we place this idealized device in the back of a pickup truck and drive it around while watching to see how the ball moves. If we drive on a smooth road, the ball rolls around but is likely to stay near equilibrium position A. Forces applied to the ball on a smooth road are not sufficient to give the ball the energy needed to overcome the boundary at E.

Since the ball is the only dynamic component, this system has just six degrees of freedom (three rotational and three translational rigid-body modes). These determine the eigenstates of this system. The track acts to constrain ball movements, and as such it limits the accessibility of the system to just three of its six eigenstates, one rotational and two translational.[8] These two modes enable ball movement about point A. The system has five equilibrium positions for the ball to settle; these are shown in Figure 6.4 as points A-E. Four points (A-D) yield stable equilibria while point E is unstable.

Eigenvalues of the system in Figure 6.4 vary in time. Together they describe uncoupled track-ball-interaction properties and determine ball

---

[8]It could be that one, two, or all three rotational modes are accessible depending on how the ball is connected to the track. Here we assume the ball is constrained so that only one rotational mode is accessible.

locations in gravity wells. Eigenvalues describe the energy required to transition the ball from its current equilibrium point to another. Accessible eigenstates are largest when the ball is located near a stable equilibrium point and smallest near unstable point E. System eigenvectors are related to the possible paths the ball can take, and these are clearly limited in this example. With roughly three degrees of freedom, this system is homogeneous.

At point E, the accessible eigenvalues go to zero, in which case the system achieves either a *bifurcation point* (Section 4.7) or opens a *rigid-body mode* (Section 4.9). A bifurcation point occurs when an eigenvalue approaches zero because of a change within the system itself. A rigid body mode becomes available when an eigenvalue approaches zero because of a change in environmental boundary conditions, but the system has not changed.[9] The latter situation exists at point E in the systems of Figure 6.4.

To illustrate, imagine we take the pickup truck onto a bumpy road and drive fast. The forces applied to the ball, which you will recall must stay on the ideal track, could give the ball enough energy to roll past point E. If that occurs, then the eigenvalues for the accessible eigenstates fall to zero and a rigid-body mode opens allowing the ball to fall toward one of the four stable equilibrium positions A-D. The path taken (including the unlikely fifth state where the ball remains perched at point E) depends on the net forces on the ball. At the instant the ball reaches point E, all six eigenstates are accessible, but just for that instant, until say the ball heads toward point D. Now the eigenvalues for the three previously accessible eigenstates become large and inaccessible, and three new eigenstates become small and accessible. The new eigenstates are a different translational-rotational combination from the six possible modes. All this analysis leads us to the conclusion that the ball has changed direction, which seems like a lot of analysis for a simple system. Yet the same analysis also describes what occurs in much larger systems that are much less transparent.

Rigid-body transitions can only occur at critical[10] point E in this example. If we know everything about the system, including all the

---

[9]Thinking back to the explosive building demolition example of Chapter 4, the explosion changed a boundary condition for the building that opened a rigid body mode, causing the building to fall. The force on the building after the fall sets eigenvalues for global eigenstates to zero. This changes the equilibrium state of the materials from "building" to "rubble".

[10]Critical values are *boundary lines* in a phase portrait for a linear (or linearized) system model. For example, the imaginary axis in a plane defining complex roots of the ODE solutions separates two spaces, where system equilibrium values are stable (negative half plane) or unstable (positive half plane). Boundary lines become *boundary manifolds* in a phase portrait of a nonlinear dynamical system model.

forces acting on the ball at that time, we could predict the post-bumpy-road equilibrium state exactly from a mathematical model. Our incomplete knowledge of the state of the system reduces our ability to use the model to predict every response accurately, giving what appears as model variability when in fact no intrinsic system variability truly exists.

Even without full information about the system, the model helps us understand which responses are generally possible. As long as system eigenvalues are not approaching zero, the ball will return to the closest stable equilibrium point. System responses to pickup-truck input forces are therefore limited to parabolic movements about stable points A-D. Also, knowing a little freshman physics, we can predict the energy required to change the equilibrium state.

A change of equilibrium state is often called a *phase transition.* In condensed-matter physics, as the temperature of liquid water slowly falls, water molecules gradually slow their thermal dance until the temperature reaches a critical value at 0^oC where the liquid transitions into a solid (Fig. 3.3). The process begins with localized crystals that grow and merge together until a solid mass is formed. At the freezing point of water, statistical models show that an entirely new range of state solutions becomes available to this homogeneous system composed of many identical water molecules, and for a time some solutions exist simultaneously. There are obvious parallels between the homogenous systems of pure water freezing and the track-ball example of Figure 6.4.

The capacity of a homogeneous system for storing and expressing information is quite small. While there are an infinite number of positional states the ball can assume, only a few paths are possible and all of them can be modeled mathematically.

Heterogeneous-system example. How might the track-ball device be modified to find a system with a hierarchy of emergent properties that is characteristic of a complex system? The answer is to add degrees of freedom that yield additional distinct response states (increased information capacity), and one way to do that is illustrated in Figure 6.5. Instead of four smooth parabolic tracks, we have one jagged parabolic pit filled with balls. We cap the ball pit so that none can fall out except for the smallest ball at the bottom whose size allows it to pass through the one hole on top.

Now we place this device in our pickup truck and head for rough terrain where the balls jump and bounce with truck movement until eventually the small ball pops out.

Fig. 6.5   One possible extension of the homogeneous track-ball system shown in Fig. 6.4 is to form the complex system illustrated in this figure. A jagged, rigid-material, parabolic-shaped ball pit that is filled with balls of different sizes is shown at rest. The pit is capped except for a hole on top that is just large enough for only the smallest ball at the bottom to exit.

The jagged ball-pit surface adds constraints to now 3-D ball locations and therefore many more equilibrium states. However, if we only had the smallest ball in the pit, we would have at most six degrees of freedom. Even with all eigenstates accessible, the single-ball system could still be homogeneous. Adding more balls to the system adds new degrees of freedom. With each additional ball, the system adds complexity and new emergent properties.

When variability increases dimensionality, a broad spatial hierarchy of eigenvectors develops. Inputs to this system are forces from the truck on the small ball imposed by driving on a bumpy road. The output is the position of the smallest ball. If one goes off-roading long enough, an irreversible transition will occur as the smallest ball pops out.

With N balls, we have at least $6N$ accessible eigenstates. Off-roading adds enough energy to the system to be sure some eigenvalues are going to zero all the time as truck motion keeps the balls in constant internal motion. Nevertheless, this system is fairly robust against the rigid-body transition occurring as the smallest ball leaves the truck because most system eigenvectors are local. If we had detailed knowledge of system states at each instant of time (the location and forces acting on each ball), we could develop a model that predicts the transition. However, given the very large number of equilibrium states (high information capacity),[11] the amount of information we would need to obtain to accurately predict a smallest-ball transition is truly astronomical! Accurate mathematical modeling of this system is improbable.

From the initial state shown in Figure 6.5, we could numerically model the experiment with Discrete Element Methods (DEM described in Chapter 4). Notice the similarities with agent-based and other numerical-modeling approaches. However, unlike current agent-based models, DEM is amenable to eigen-analysis because all relevant state information can be recorded as a matrix during the course of the simulation. What is missing in other numerical modeling approaches is a systematic procedure for efficiently measuring eigenstates from model data.

Transitions in the homogeneous fox-rabbit system are analogous to the track-ball system. The homogeneous population model with only a few degrees of freedom has state transitions predicted by the accessible eigenvalues, each composed of model parameters. As real parts of the eigenvalues approach zero, the solution space suddenly opens to add and remove accessibility to equilibrium states, just as new track segments opened in the ball-track system. In principle, homogeneous population behavior is mathematically predictable ([Murray (2002)], Appendix A). Transitions in homogeneous systems are openings and closings of accessibility to regions in solution space that makes previously impossible responses possible.

Transitions are predictable from mathematical models of homogeneous systems as bifurcations and rigid-body modes. One of the differences between the homogeneous predator-prey example and the track-ball or loaded-column examples is that animal populations can transition from a sustainably oscillating equilibrium state to a steady equilibrium state and vice versa. These transitions are prevalent in linearized models of nonlinear systems that include time delays. You can imagine many realistic delays that might be included in the model, like the lags between the time a predator is born and when it is mature enough to hunt rabbits. Transitions from oscillating to constant equilibrium states and back are referred to as *Hopf bifurcations*. Of course, we do not see transitions as clearly when they occur in realistic complex ecosystems because of the analytical obfuscation

---

[11]Generally adding variability to a system increases entropy and decreases information. However adding variability as described here increases information *capacity* in the Boltzmann sense.

wrought by high dimensionality. We don't fully understand the role of time delays in complex system behavior; it is important and worthy of a closer look.

Summary. At the moment a complex system transitions from one equilibrium state to another, the degrees of freedom change, sometimes temporarily, and rigid-body modes can form as eigenvalues approach zero causing system components to respond coherently in directions dictated by the corresponding eigenvectors. A clear example of these effects in a complex system was illustrated by building demolition using explosive charges in Figure 4.7.

In homogeneous systems, there might be many eigenstates but only a few are accessible to the system for generating responses. The invariance of homogeneous system properties means the accessible modes are mainly global in their extent over the system domain. Transitions in homogeneous systems are often obvious to observe and predictable through modeling. At the onset of a transition, a rigid-body mode opens or a bifurcation point is reached. Mathematically speaking, alternative solutions suddenly become available to add new system responses. Homogeneous-system properties before and after a transition are determined by scale-invariant modes.

In heterogeneous systems, however, many more of the intrinsic eigenstates are made accessible because of the diversity in properties that exists among system components. Inputs to a complex system will generate instantaneous responses predicted by linear combinations of uncoupled system properties that are manifest at each instant of time. The many degrees of freedom generate accessible eigenvectors that influence system behavior over a scale-varying range from local to global. The spatial hierarchy of modal influences generates different emergent properties at different spatial scales. Since observable responses of a complex system undergoing a transition depend on accessible-eigenvector scale, transitions are difficult to predict mathematically unless we learn to efficiently perform eigen-analysis on computational models of complex systems, or if we can identify vulnerable eigenstate indicators. Agent-based modeling of complex systems has much analytical promise. The good news about spatially-varying properties is their ability to buffer against transitions in ways not available to homogeneous systems.

6.7 Inaccessible Dimensions and the Influence of Subspaces

We saw that a high-dimensional system can be arranged so that most of its eigenvalues are set much larger than the rest. Recall that we formulate the analysis so it generates eigenvalues that are monotonic with the energy required to activate the corresponding eigenvectors in a system. Even if a system has many eigenstates, it will behave like a homogeneous system because its properties are determined by the *accessible eigenstates*, those few with the smallest eigenvalues. These high-dimensional systems behave like low-dimensional systems with little component diversity.

Fig. 6.6 Collections of identical rigid spheres on a flat table (left) and variable stones on a rocky beach (right).

The consequences of limited eigenstate accessibility can be illustrated using the mechanical system of Figure 6.6. But first, we wish to point out that there are many analogous situations in large systems where this occurs. For example, in biological gene expression, inactive modes are selectively activated and deactivated by supplying the cell with appropriate signals. Cell systems use lock-in-key signals in place of energy barriers to switch on and off many hibernating functional modes. Readers are urged to draw parallels between the example below and systems you know.

We examine the two collections of components illustrated in the photographs of Figure 6.6. On the left, identical rigid spheres are shown placed on a flat, low-friction, horizontal table surface. The idealized sphere-table system can be said to be homogeneous because the ideal spheres are identical and the ideal table surface is uniformly flat and featureless. In contrast, on the right side of Figure 6.6, we see a beach with stones having highly heterogeneous topological properties. With different shapes, the stones lay in soft sand or on each other; beach component properties are highly variable.

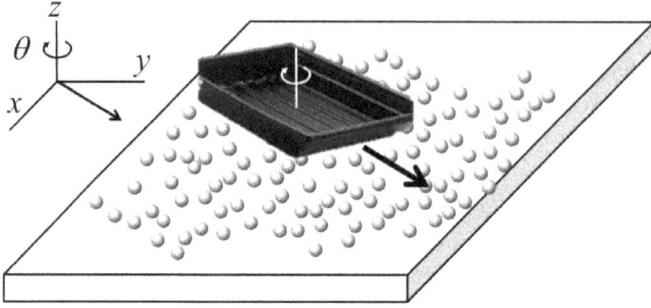

Fig. 6.7   A homogeneous sled-spheres-table system is illustrated. The smallest three eigenstates correspond to translation in the x, y plane as the sled rotates at angular velocity θ (degrees/second) about the z axis.

Imagine that a rigid flat-bottom sled is placed first on the spheres and pulled along the tabletop, as illustrated in Figure 6.7). The homogeneity of identical spheres enables the sled to glide long distances on the nearly frictionless surface with just a tiny pulling force. Pulling-force energy is the input stimulus and sled displacement is the output response of the system. If the weight of the sled is spread over a larger area, we just add more identical spheres to achieve the same output response for a given input force. The scale-invariant properties of the sphere-sled-table combination are determined mostly by properties of the components (sphere size and density), as expected for a homogeneous system.

Examining the eigenstates, we can predict the responses of this system to an applied lateral force, and in this example the analysis is straightforward. Sled movement is described by the eigenvalues of the six rigid-body modes of sphere motion. The accessible modes have eigenvalues very close to zero, and because these eigenvalues are nearly equal, their eigenvectors may form a subspace.[12] In some situations, energy from one accessible mode is free to transfer to another accessible mode. Therefore, any small input force moves the sled anywhere in the plane, and in some cases the sled will rotate as it translates (fishtailing). Eigenvectors for the other three rigid-body modes all involve movement along the z-axis. Since those eigenvalues are larger, substantially more energy is required to lift or tip the sled out of the plane of the table than to glide it along the table.

---

[12]Note that if three eigenstates are accessible with equal eigenvalues, there can be less than three degrees of freedom. Please realize there are many details not specified here.

Homogeneous systems like this one have low information capacity. That is, sitting on the sled the only sensation you will note is a change in sled acceleration. Without landmarks, the sled's position, direction and speed are impossible to know. The low dimensionality of the homogeneous sled system enables us to easily predict its responses to stimuli using mathematical models even without sensory data from motion sensors or cameras.

Moving the sled from the spheres to the beach stones and pulling, we immediately notice a big difference in response. The natural variability in stone size, shape, surface roughness, and relative height in the sand means the sled now resists any movements from a tiny force. Pulling harder until the sled moves, we find it bobbing up and down as it slides across the rough beach surface. Like the spheres, each stone has six degrees of freedom if we ignore deformability, but, unlike the spheres, stone sizes and shapes are unique and asymmetric. The primary interaction between stones and the sled is frictional. Unique component properties mean the sled's movement patterns change continuously as it moves and experiences different stones beneath it. The instantaneous combination of stone-sled properties determines the sled's bouncing and bucking movements across the beach. Any movement requires a large applied force likely to activate all of the eigenvectors representing each stone as it contacts the sled.

If we change the contact area of the sled on the beach stones, even without changing its weight, we find the movement patterns also change in a way that is unique to each stones the sled encounters. These are the scale-dependent properties expected of a complex system. The high-dimensional sled-stone system has emergent properties that continuously change with sled movement. Even though the eigenvalues for some local eigenvectors of the system may fall to zero, the overall direction of sled motion remains about the same suggesting robust global properties. The "randomness" of motion is not a fundamental property of the system; rather it is a product of our ignorance of details about the state of the system at each moment in time.

It would be very difficult to develop an accurate mathematical model to predict the responses of the sled-stone system to an applied force. The high information capacity of this high-dimensional system means we need to measure an extraordinary number of constantly-changing properties of the sled as it moves. Accurate mathematical modeling of this complex system is theoretically possible, but highly impractical and prohibitively expensive. There is so much information contained in this system that you could determine your position, direction, and speed if you were riding along just from sled movement provided you knew everything about each component property of the beach and built a model with this information. In this way, the sled-stone system is functioning as a *probe* for sensing some of the beach properties.

System probes.   Note that the beach with a sled is a different system than one without a sled. A sled is a system component with constant and knowable properties. Therefore, applying a known input force to the sled generates sled-position responses we can measure and use to discover beach properties. If we place instruments on the sled that sense its position, speed, etc., the sled becomes a probe for exploring surface mechanical properties. Probes detect properties that may not be otherwise observed. They are effective in this role provided their use does not significantly alter the system being sensed. For example, if the sled was heavy enough to plow a channel into the beach, then its sliding properties are irreversibly changed as a result of the measurements. If we are interested in other beach properties, like temperature variations over time, we need to change the probe so it senses thermal properties without altering them.

Monitoring probe activity within a complex system noninvasively is a well-known technique in many fields, especially in biology and medicine where invasive assessments are accompanied by serious risk of injury. For example, *molecular probes* can be used to assess molecular functions in the human body, noninvasively. These probes are composite structures composed of a chemically active *carrier* molecule naturally found in an organism that is attached to a *reporter* molecule. Injecting functionally active molecular probes into body's blood circulation at low concentration, we can track their natural movement using imaging technologies to reveals the location and proximity of molecular probes in and around cells. If the probe can access all relevant body compartments without altering the physiological state of the patient and be detected, we obtain vital information about diseases and patient health. This is the basis of molecular imaging, a rapidly-developing tool available to 21st century medicine.

Central control, energy and information.   Adding component variability increases the *information capacity* of the system as the number of potential response states expands. The number of positions that a sled can assume on a rocky beach are clearly much larger than the number possible when it is placed on the spheres because, in the former, both vertical and horizontal movements are available to the system. Information capacity describes the number of states that a system can assume.

The heterogeneous beach system takes little energy to create because it is found naturally but much energy is required to operate a moving sled. In contrast, a homogeneous system requires substantial energy to create and maintain but little energy to operate. The skill and energy that goes into manufacturing hundreds of identical spheres and a flat tabletop is only possible when *central control* is applied from sources outside of the system. We could machine and polish stones into spheres and level and pave the beach sand flat and hard. In doing so, we would eliminate most variability and convert a complex system into a functionally-homogeneous system by changing the accessibility of its eigenstates.

Conversely, the second law of thermodynamics ensures that homogeneous systems become more heterogeneous over time – that is, unless we provide the energy necessary to maintain the homogeneous state. Sliding a heavy sled over glass spheres would eventually increase the variance among sphere-table properties. Grooves will form in the table and the spheres will crack and crumble. The tendency of entropy to increase naturally drives all systems toward complexity, even cell systems that are unique for their ability to add information to their already high information-capacity selves. Complex systems are naturally found throughout the universe wherever we look. Homogeneous systems are unnatural and manmade. They are most often encountered when we represent a real system in a simple model as we attempt to predict and ultimately understand some aspects of its behavior.

Thus far, Chapter 6 describes *thought experiments* involving physical systems to explain the fundamental role of variability in determining the properties of natural and manmade systems. In the remaining part of the chapter, we use examples from other fields of study to suggest these concepts apply generally.

6.8 Social Variability and Stability

Let's switch from mechanical and population modeling examples to business and social system examples. We should be able to do this if the same principles apply across all systems.

There is an oft-mentioned observation that productive organizations actively structure staff recruitment to obtain a workforce with a wide variety of educational and employment experiences. From the eigenview, we say they hope to increase the number of degrees of freedom in their enterprise by increasing component variability. Such organizations value adaptivity and creativity in work product more than the security of predictability and uniformity; consequently, they aim to increase organizational heterogeneity to broaden the spectrum of emergent properties.

Fig. 6.8   Successful collaboration requires parties to share goals, knowledge, and respect. These attractive forces are required to balance the repellent forces exerted by creative individuality within a diverse workforce.

At the same time, all successful and creative collaborations in any human enterprise must nurture attractive forces between the component parties. These forces bind long-term working relationships. They are based on there being *mutual goals, knowledge, and respect* among the interacting parties. These features establish the component connections that make it possible for an organization to form a productive complex system. The best organizations balance the cooperative order with the creative productivity that are both required for success. They do this by managing internal variability.

Intellectual and emotional variability are also essential for some organizations to function at all. The most obvious examples are stock markets. They achieve an equilibrium state only when investors think differently from one another. Trades on a company stock depend on some traders thinking a stock is overvalued and are willing to sell it while others are thinking it is undervalued and are willing to buy. Variability is existential for this system. If all investors thought exactly the same way at the same time about stock values, there would be no trades and the markets would collapse.

Variability in human behavior is responsible for the emergent properties we observe in all socioeconomic systems. Variations in the backgrounds and abilities of adaptable people enable society to fill all the jobs necessary for it to function. Some people prefer to live in high-rise condominium buildings at the center of a city while others prefer to live in houses in suburban or rural areas. Some drive to work, while others use public transportation, bike or walk. Societies exist in stable but dynamic and evolving equilibrium states because of (not in spite of) the variability that causes individuals to think and act differently. Diversity can destabilize organizational equilibria if we allow our base nature to control rational thought. However, at our most tolerant and welcoming best, it is our differences that enrich daily interactions to add value to our lives. We also hear that countries with diverse populations that are free to express their individuality are found most productive long term as measured using quality-of-life metrics. Yet these features make society vulnerable to civil unrest as subcultures clash when the binding forces among people (mutual goals, shared knowledge, and mutual respect) break down and are lost.

If diverse component properties are required for complex systems to exist, what happens to the stability of a functioning equilibrium state when variability is suddenly lost? For example, what if there was no variability within the flock of birds described in Chapter 1? The short answer is that the flocking property would soon disappear. Imagine that suddenly, with a magic wand, we removed all variability among prey birds; now all the birds are exactly the same. We mean "exactly" the same; their age, health, physical ability and biological make up are exactly the same. We saw in Chapter 1 that the predators are best able to isolate and capture weak birds that may have become old or sick. With homogeneous populations, the predators find no slower prey to capture. The flock would achieve a perfect flocking strategy that evades attack and protects "all" the birds within the flock, and predators would soon give up hunting this prey. With the disappearance of a major source of nutrition, there could be a reduction in the predator bird population leading possibly to extinction. Without a predator threat, flocking behavior in prey populations would no longer serve any purpose, and that behavior would gradually disappear. Clearly variability is essential in both forming and maintaining system properties.

Sometimes variability is to be minimized when the overall system goals are predictability and uniformity in output responses. If the handling of your car changed every time your drove it, you would be very unhappy with that product. Effective automobiles place central control in the hands of competent drivers. Since drivers vary in skill and attention span over time, increasingly automobile manufacturers augment driver control with automatic braking, lane-changing indicators, and parallel parking functions that enforce central control when drivers perform poorly. Autonomous components are intended to add more reliable central control to automobiles by minimizing driver variability in key situations.

6.9   Variability in Biological Systems

At various points throughout the book we discussed the idea that organisms form a unique type of complex systems. Their extraordinary degree of variability at different scales qualifies living systems as among the most complex we know, in every sense of the word. The source of a cell's special properties is a shared memory with other cells in the body via the DNA molecule. DNA provides the instructions needed to build molecular order from environmental disorder. Cells begin the process after receiving a signal calling them to action by following their genetic instructions to assemble protein-based molecular manufacturing machines. A cell builds all the machines it needs to then build the parts necessary to meet its metabolic needs, repair damage, communicate with other cells, change its core functions, and copy itself. The products of cells creating order from disorder are multicellular organisms with ensemble properties we label anatomy, physiology, and personality. Their manufacturing mission identifies most of the cells in our bodies as information machines. What fundamental role does variability play in these systems?

Let's address this multifaceted question by first focusing on *meiotic cell formation*. It will help to illustrate how organism development blends aspects of increasing and decreasing variance. Meiosis is a process that forms cells in animals and plants that can later combine with a cell from another member of the same species[13] during sexual reproduction to create a new multicellular organism. *Diploid* cells (cells with the usual two sets of chromosomes) will assemble duplicate copies of their chromosomal DNA to have four sets of chromosomes for a short time. Some of that material is rearranged (e.g., chromosomal crossover) in each set before that cell divides twice in a way that is different from other cells in the body. The four new

---

[13] A species includes individuals capable of reproducing or exchanging genes. A measure of the "success" of a species is its population growth rate.

haploid (one set of chromosomes each) daughter cells each have one half of that parents chromosomes, but crossover adds variability to the parental information to be passed to progeny. If the rearrangement is viable, these cells can mature into gametes, spores, pollen, and other reproductive cells depending on the species. In human reproduction, gametes from two parents (egg and sperm) fuse to form a *zygote* that combines the parental genes in a manner that adds more variability. The DNA passed from parents to the child has a very similar sequence order, so the child inherits many parental traits. However, meiosis introduces enough variability to the genome of the nascent zygote, now dividing to form a *blastocyst* and eventually a *fetus*, to always generate a distinct organism.

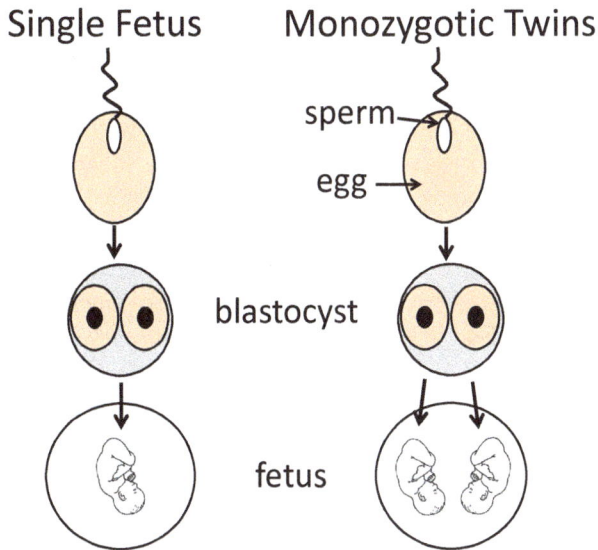

Fig. 6.9   Single and identical twin human fetuses.

We are each genetically unique; even *monozygotic* (identical) twins are not strictly identical (Fig. 6.9). Monozygotic twins form soon after conception when the blastocyst splits completely into two parts, each with the same copy of the genetic code. Even if the split occurs at the two-cell stage, there can still be genetic variability between the developing fetuses because of *copy number variations* that occur in more than 10% of human DNA. In some DNA regions there can be unusual numbers of copies of a particular gene, either too many or too few. Additional copies can add up to a million base pairs in just one part of a three-billion base-pair DNA molecule. Hence identical twins with copy number variations may have different traits and in some cases the differences can increase the risk of disease. With more than 20,000 genes in each human, the chance for copy number variations between twins is significant. The bottom line for this discussion is that procreation is a process that *increases variability* among the population of a species.

After conception, the blastocyst proceeds with its preprogrammed mission of building exact copies of the molecular machines necessary to faithfully proliferate via a different process known as *mitosis*. Cells within the growing lump of cells will normally specialize and cluster to form tissues within the developing organism through a combination of cellular division, differentiation, motility, and morphogenesis. As cells are added, details of the original cell's unique genetic code are amplified to form emergent functions at different tissue scales. Emergent properties are often seen at the largest spatial scales. For example, a pathologist would tell you that we all look pretty much the same under a microscope, while we know that our appearance and personalities are much more unique.

Neonates continue to develop after birth by responding to environmental inputs (biological, emotional, social, etc.), which further amplify or attenuate the inherited traits. Nature-nurture forces mold individuals and ultimately the species. The bottom line for this discussion is that environmental forces tend to act by *reducing variability* within the population of a species.

Developmental amplification and attenuation of inherited genetic variations in the context of the environment form scale-varying emergent properties we recognize as intelligence, health, and personality. In the course of an organism leveraging its unique properties during a lifetime, s/he adds variability to the environment even though the environment (via society) acts to reduce variability within the population. Successful individuals, i.e., those who successfully balance these opposing forces, are rewarded with an opportunity to pass on their genes biologically and their accumulated knowledge personally.

The development of a successful business is similar to that of a successful organism. The best businesses locate and create environments that balance variance-reducing attractive forces, leading to collaborative collegiality and friendships, with variance-increasing repellant forces, leading to competition that rewards creative expression and advancement. The optimal balance of these forces leading to financial success and influence depends on the business goals, the imposed environment, and timing. We see this model of business development as a manifestation of natural biological processes, the same ones that bring nascent life forms into environments where they interact to develop, excel, or not. It seems that every aspect of successful living, human and otherwise, involves balancing forces working to increase and decrease variance. Organisms and the organizations they create are complex systems that operate following similar basic principles.

Imagine a hypothetical world in which there is no variability among humans. We would have to return to a variation of the absurd situation of Figure 6.1 where homogeneous fox and rabbit populations interact in a homogeneous environment. Waving the magic wand once again, we would find people with exactly the same features and mannerisms constantly experiencing the same situations simultaneously. It is simply impossible for such a world to occur anywhere in the universe we see around us. Just think how difficult and unrealistic it would be when one of the authors goes to breakfast with his wife on Tuesday mornings to find everyone else, including the restaurant employees, have the same idea and there is no one to make and serve breakfast!

6.10 Variability in Economic Systems

We pointed out that homogeneous systems exist mostly in our minds as models for analyzing the complex systems found everywhere in nature. We can use mathematical equations to predict all aspects of system behavior, including transitions occurring at bifurcation points. Unfortunately, no systems found in nature are accurately predicted by these models; they are best used to understand qualitative features of system behavior but their predictability is limited. So why don't we just break up large complex systems into small parts where the analysis is already available? The answer is simple. If we extract a small part of a large system for analysis, the lower-dimensionality of the abstracted model may be missing the emergent properties that make natural systems so rich in behavior. We need to study "whole systems" when possible, and for this we must discover how to perform eigen-analysis on all manner of computational models. In this way we might find it possible to model and quantitatively analyze large complex systems ... maybe ... we hope.

Such tools might also help predict the behavior of systems with variable numbers of degrees of freedom that change over time. For example, stock markets in 2014 seem to function in reasonably-stable dynamic equilibrium states, increasing steadily despite a volatility considered by many investors to be quite high. Since 1993, the volatility index or VIX was proposed to measure the percent deviation of market indicators from expected values. In 2014, the VIX was around 10-20%. High volatility is tough on individual investors and businesses but it offers overall market benefits. It eliminates marginal firms and makes it harder for all start-ups to survive. Despite these apparent negatives, market volatility also works to redistribute capital into uses that strengthen the markets overall, much in the same way that evolutionary forces can be hard on individuals but work over time to strengthen populations.

We see variability in societal needs driving up diversity in new and existing businesses while diversity is being reduced by market pruning. It is not surprising that markets thrive once they settle into a natural balance between all of the forces that increase and decrease variability. After all, markets are driven by biological systems whose evolutionary path was guided by an analogous balance among forces.

Variability in component (corporation and investor) behavior expands the price range of the stocks being traded, which is healthy for the market. That variability disappears if some unfavorable information about a company becomes public. The price range of that stock can suddenly narrow to sharply lower values just as we find public opinion becoming uniformly low. This is an example of a rigid-body mode opening after a transition triggered by a loss of variability has occurred. The loss of variability for this one stock price can be an isolated event without affecting the whole market. In some cases, however, if the stock is widely traded, and, as company stock prices uniformly plunge, a few eigenvalues for the whole stock market may approach zero and it may initiate a global market transition. Market-system transitions can also occur in the opposite direction when favorable information about a company becomes public. There is great benefit for a few investors but the overall effects on the market are often detrimental. Except for the most radical libertarians, most would agree that to keep free markets healthy there must be some regulation, i.e., central control from outside the market system.

It is interesting that inputting reliable positive/negative information into the market system can transition the whole market with the effect being that investors will eventually act to buy/sell the affected stocks. If the stock is blue chip, meaning its value is a market indicator making the associated eigenvectors highly distributed, the transition can greatly influence overall investor confidence and market performance. Otherwise, the effects of the transition will be localized without broad consequence. Clearly the stock market is a very special high-dimensional complex system, quite sensitive to the slightest insider information, and thus highly vulnerable to sudden transitions causing damage to its volatile currency known as *investor confidence*.

Reductions in investor variability can cause a bear market leading to recession or a bull market leading to a bubble where prices are driven above their objective value. Either can happen in a number of ways. When the stock market is in equilibrium there is sufficient variability in public confidence with groups of people thinking the economy is about to improve while others feel it will stay the same or decline. When suddenly almost everyone thinks that the economy is likely to go into recession, the decline in attitude variability causes some economic-system eigenvalues to approach zero. At this point in time, the stock market is ready to transition into a bear market at the slightest trigger, and when market volume is high there can be many triggers. The opposite transition may occur when almost everyone thinks that positive factors are affecting the value of stocks and will continue to do so for the foreseeable future. The reduction in variability in the latter case can be gradual and it may cause the stock values to increase to unreasonably high values causing a bubble. As the bubble increases, some of market eigenvalues approach zero, making it increasingly vulnerable to a sudden collapse.

This is what happened to the US economy during the 1997 – 2000 period in what is known as the "dotcom bubble." The variability in confidence towards growth of the internet sector and related technology stocks gradually decreased. Most investors started believing that the rapid growth in these stocks was likely to continue. Alan Greenspan, the Federal Reserve Bank chairman at that time, famously called it "irrational exuberance." As stock values soared to unreasonably high values, it seems likely that the real part of some market-system eigenvalues approached zero, although we cannot now validate that statement. The market became increasingly vulnerable to a sudden transition that triggered a bear market with rapid declines in stock values. This transition started in March 2000 when the tech-heavy NASDAQ index peaked at 5046 (Fig. 6.10). Over the next 2.5 years many new technology companies went out of business and the NASDAQ lost 78% of its value to close at a low of 1114 during October 2002. The dotcom bubble event is an example where narrowing system variability, often referred to as "herd behavior," led to market instability and eventually an equilibrium-state transition.

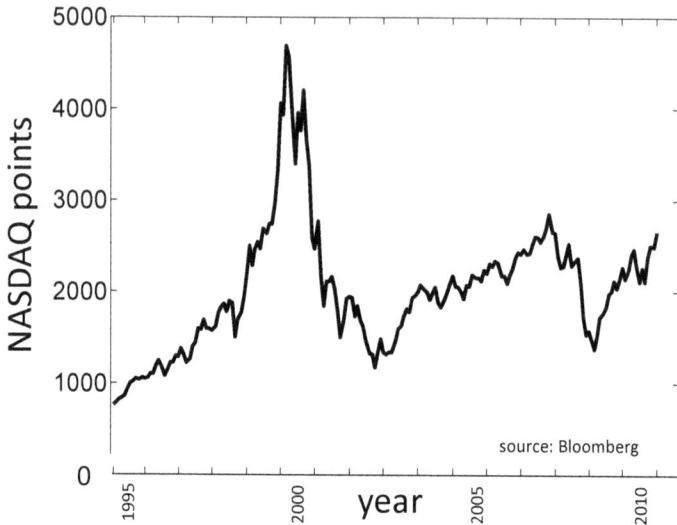

Fig. 6.10   On March 10, 2000, the NASDAQ index of leading technology shares spiked. It is believed that this event indicated the bursting of the "dotcom bubble."

Other bubble-burst transitions have occurred throughout history. It is generally believed that the first major economic bubble burst was the "Tulip Mania" of 1634 – 1637. Tulips were considered exotic at the time in parts of Europe. Their unusual beauty and hardy nature placed them in high demand, and high demand built a futures market where retailers vied for the latest cultivar. Widespread belief about continuing demand for bulbs drove prices above their value in what is today The Netherlands. This activity was quickly followed by a market crash allegedly triggered by trade alterations caused by the bubonic plague (Fig. 6.11).

Other examples include: widespread speculation in stocks culminating in the stock market crash of 1929 that led to a global depression; Japanese real state and stock market bubble collapse of 1980s that led to an extended period of recession and deflation; the 2008 real-estate bubble in the US and most of the industrial world that caused a financial system collapse. It was triggered by the bankruptcy of Lehman Brothers Bank in September 2008 leading to the "Great Recession".

Reducing variability can also lead to transitions in biologically-influenced economic systems; for example, the Great Irish Famine of the 19th century. Spanish explorers introduced potatoes into Europe during the

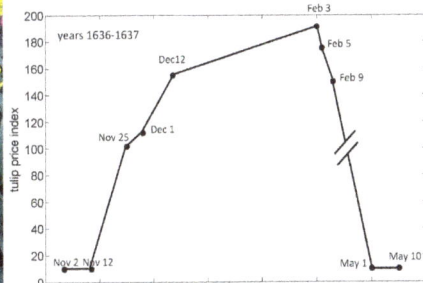

Fig. 6.11   The price index for tulip bulb contracts formed an economic bubble during the seventeenth century. Data from [Thompson (2007)].

second half of the 16th century where their use rapidly spread. The particular species of potato introduced into Europe lacked the genetic diversity found in many indigenous species found in South America, and low genetic diversity made these plants vulnerable to diseases. Failures of potato crops happened many times in Europe with varying degrees of severity and with equally variable impact on socioeconomic systems.

The most severe and consequential crop failure took place in Ireland in 1845. At the time, the Irish population was growing much faster than that of Europe (Fig. 6.12), and about a third of the Irish population depended on potatoes for food. During the 1840s a fungus-like blight devastated the potato crop over just a few years. Crop failures coupled with harsh politically-motivated Irish laws created a famine during 1845 – 1852 that led to a million deaths and mass emigration over the following century. Adopting the eigenview, it appears the growing dependence of a rapidly expanding population on a single food source narrowed variability in this system, making it vulnerable to a transition. Coupling this event with politics of the time seems to have amplified the devastating effects.

Biologists attribute the crop failure to the lack of genetic variability in the plant species. Then and now we find there are hundreds of potato species. Many that are cultivated in the Andes Mountains have genetic

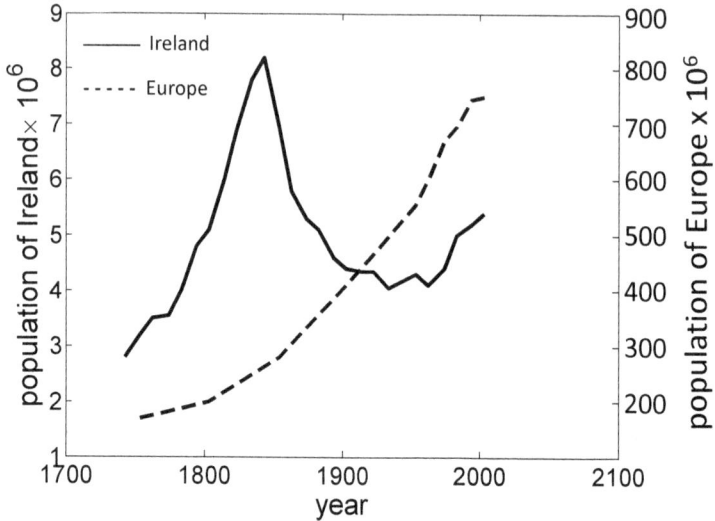

Fig. 6.12   The populations of Ireland and Europe during the Great Famine years.

diversity that offers plant resilience and stability from blight but perhaps a few culinary shortcomings.

We have not looked closely at the economic, political, cultural, and agricultural components at play in Ireland at this time in history, but we nevertheless are struck by the resemblance of the curves in Figures 6.10 – 6.12, where in these three disparate examples we find that a loss of component variability led to system instability and transition. If we had a way to measure the eigenstates of these systems, we are convinced we would find the extent of the transitional impact would depend on the distribution of eigenvector weights for the eigenvalues that approached zero.

In each example, a loss of variability reduced the number of accessible eigenstates. States were lost as eigenvalues fell to zero and rigid-body modes opened causing the affected components to suddenly respond in a similar manner. Some might believe transitions begin when the bubble bursts. The eigenview of the situation identifies the transition beginning with the formation of the bubble.

6.11   Transition Susceptibility in Ecosystems

Transitioning ecosystems are among the most disruptive on Earth as discussed in Chapter 1. As agriculture came to realize, variability plays an important role in determining the susceptibility of these and all systems to sudden transitions. Ecologists use the terms *resilience* and *stability* when describing equilibrium states of ecosystems. They define resilience as a measure of how quickly systems recover from input perturbations such as storms, floods, fires, human activities, or the introduction of invasive species. Resilient ecosystems return nearly to their previous equilibrium states quickly. Stability is defined as a measure of ecosystem constancy; essentially the resistance to change [Gunderson *et al.* (2010)]. The two terms offer different but related information about an equilibrium state. A highly resilient system is unstable if it is easy to perturb the system from its equilibrium state but it readily returns to equilibrium when perturbations occur.

Even highly stable and highly resilient ecosystems can transition. The term "regime shift" in ecology is similar to equilibrium transition used to describe other complex systems. We do not currently have the capability to determine, or even define, the eigenvalues of an ecosystem. Ecosystems are nevertheless complex systems with eigenvalues describing uncoupled properties. Reducing variability/diversity in ecosystems reduces the number of accessible eigenstates, which may activate rigid-body modes thus facilitating often irreversible transitions. Human interference in the management of eco-socioeconomic systems has led to reductions in biodiversity, which has been extensively documented in the fields of ecology and environmental sciences [Holling (1996)].

One case involves the reduction in diversity of trees in forests of eastern Canada. Holling noted that spraying trees to reduce the spruce budworm population delayed existing tree mortality, which was good for the pulp mills in the short term. However, the practice reduced tree diversity over the region, and with time a consequence of the more homogeneous system became greater vulnerability to intense outbreaks of infestations. Long term, these events generated high tree mortality and required more monitoring and spraying than was originally required.

The second case involves reduced diversity in West Coast salmon populations. To enhance abundance for sport and commercial fishermen, hatcheries were developed along the North American west coast. As hoped, fish populations expanded quickly and with more predictable patterns, leading to more hatcheries and a larger fishing industry. However, this practice contributed to the extinction of the indigenous species, leading to reduced natural diversity in salmon populations, leaving the industry more vulnerable to sudden losses from climate change and disease.

In both cases, human intervention was intended to increase productivity to benefit the socioeconomic part of the system. These short-term successes appear to have reduced diversity in the long term, making the whole eco-socioeconomic system more vulnerable to irreversible transitions falling into undesirable equilibrium states.

6.12   Self-organized Criticality

We have been discussing how transitions in simple/homogeneous systems can be predicted by monitoring eigenvalues computed from mathematical models. These low-dimensional systems are amenable to conventional mathematical modeling techniques. Models predict that when critical values of equation parameters are reached, real-parts of eigenvalues go to zero and the expressed behavior of the system is suddenly modified to reflect overall changes in system properties. An example was illustrated in Figure 3.2, describing water molecules undergoing phase transitions at constant atmospheric pressure. In this context "phase" refers to the macroscopic equilibrium-state properties that form as a consequence of microscopic interactions among molecules. As pure water cools and the molecular kinetic energy is reduced, there is a sudden transition from the gas phase to liquid phase when the temperature reaches 100^oC. At its condensation point, water molecules break symmetry to weakly link. This transition produces a sudden reduction in molecular motion as the liquid phase forms. Cooling the liquid further slows that molecular motion in proportion until there is another sudden phase transition to the solid phase at 0^oC. Throughout the experiment, the input parameter *temperature* indicates that energy is slowly removed from the system at a constant rate. The system responds linearly to this input with a gradual reduction in the thermal motion of molecules until at specific temperatures there are sharp (nonlinear) reductions in molecular motion. (The full story is quite a bit more interesting when one considers changing both temperature and pressure.)

Homogeneous systems have transitions analogous to physical phase transitions, and hence "phase transition" has been applied broadly to describe many phenomena. As the ball of Figure 6.4 smoothly rolls along track A to reach critical point E, its direction of motion can suddenly and dramatically change in response to a tiny input force, sending the ball into a new "phase" or equilibrium state. Rolling may be considered as a slow continuous adjustment of model parameters, analogous to temperature. When the ball reaches critical value E, an eigenvalue suddenly goes to zero and the system transitions from one track state to another. *Criticality* is a term that defines the behavior patterns of a system during a transition. Mathematical models of homogeneous systems can provide quite a bit of intuition about these phase transitions.

High-dimensional complex systems behave differently. Their components properties vary significantly so that the network generates many degrees of freedom and system properties emerge with many accessible eigenstates. Component properties are reflected in the eigenvalues of complex systems and in the associated eigenvectors that spatially distribute their component weights hierarchically, generating a spectrum of local and global eigenvectors. The spatial extent of the eigenvectors associated with the zero eigenvalues of a transition is what determines the observable effects of the transition. Complex adaptive systems are difficult to model mathematically, making it difficult to monitor the eigenvalues that could predict the susceptibility of the system to transition and, perhaps, the changes in system behavior likely to result.

We have been discussing transitions as rare and extreme events, but that is not always the case. There are systems that reach a sustainable state of criticality. For example, there are combinations of water temperature and pressure that, if held constant, gives multiple phases of water simultaneously. At these critical parameters, we might think of water molecules existing between phases. Similarly, if the ball on the track happens to stop at critical point E, it remains between four stable equilibrium states; its system eigenvalues are maintained at zero value. Homogeneous systems can remain in such a critical state with a great deal of outside control but they are usually highly unstable and don't usually remain at criticality on their own.

Complex systems are much more interesting. Some self organize into a naturally-forming, persistent state of critical behavior. The general phenomenon was introduced about 30 years ago by Per Bak and colleagues as *self-organized criticality* [Bak *et al.* (1987)]. Since then, it has been applied to problems in geology, engineering, neuroscience, and economics, but perhaps the simplest introductory explanation is provided by the original simulation experiment using granular materials; see Figure 6.13.

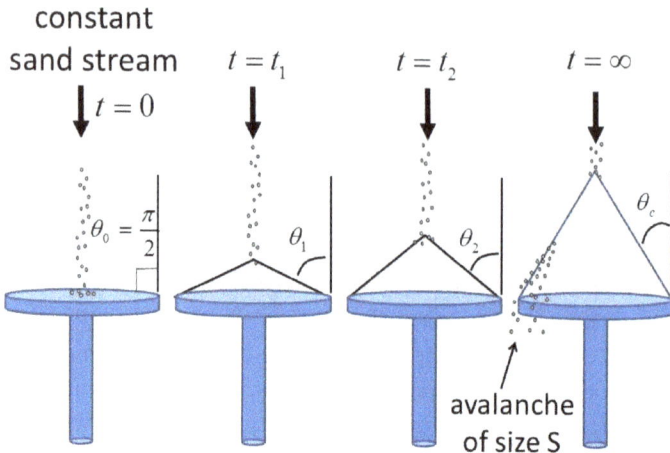

Fig. 6.13   Sand pile experiment. The mass of a sand pile grows over time as sand is added, until the slope of the pile is reduced from $\theta = \pi/2$ to the critical angle $\theta = \theta_c$. Adding more sand does not change the slope of the pile since additional mass is shed as avalanches. The value of the critical angle and the distribution of avalanche sizes S depend on properties of individual sand particles. Variability among individual particle properties is an important factor determining overall system properties.

The experiment is to pour dry beach sand onto a clean flat surface. As sand accumulates, it forms a conical shape such that the slope of the cone is reduced from 90° to the critical angle θ_c. From that point on, additional sand is shed as avalanches and the sand pile remains in a critical state for reasons discussed in the next paragraph. Avalanches might spread the sand-pile base if we provide a larger base surface, but the height of the pile also increases in proportion to the base so the slope remains at θ_c.

The dry sand pile is a complex system with component particles that interact through gravity, inter-particle contact and surface friction over long distances: in principle, over the entire pile. The shape and texture of grains determine the eigenstates of the system including the spatial weighing of the eigenvectors, which define the particle connections that determine the sizes of avalanches. The most global eigenvectors describe the overall conical shape of the pile, and those eigenvalues are large predicting a stable pile shape. The most local eigenvectors describe the smallest-scale ensemble connections among particles. When eigenvalues for localized eigenvectors go to zero, we have small avalanches, and when eigenvalues for more extensive (less localized) eigenvectors go to zero, we have large avalanches. This system structure forms a state of self-organized criticality once the pile builds and a critical angle is reached. The most trivial inputs to the system can trigger a sudden avalanche resulting in a new equilibrium state. Transitions are expected at any time because a complex system in a state of criticality has many eigenvalues very close to zero. Adding just one more sand particle might do nothing to the system or it could trigger an avalanche. Critical complex systems are able to remain critical without outside influences because of the robustness provided by the spatial hierarchy of eigenvectors.

Bak and Paczuski describe the phenomenon as inevitable transitions that are not easily stabilized [Bak and Paczuski (1995)]. They felt the future course of events would proceed through random happenstance to unfold an irreversible history of events. Of course we agree, except to point out that if the experimental course was to be repeated, so that we were able to drop the same sand particles in exactly the same way, we would observe exactly the same set of particle responses including the same pattern of avalanches. The randomness in the system is not unknowable as it is in a quantum-mechanical system. This randomness is a product of our ignorance of large-system details. With full knowledge of microscopic properties and eigen-analysis, it would be possible to exactly predict the behavior. Of course, obtaining full knowledge of the state of each particle is a tall order, so functionally we agree with Bak and Paczuski if not exactly in principle.

Let's examine the system illustrated by Figure 6.13 a little further through a series of thought experiments involving other types of granular materials. First, we use identical dry spheres with a surface treated so that there is virtually no inter-particle friction, static electricity, or any adhesive or attractive force between particles. All of the interactions between falling particles are between nearest neighbors so a pile never assembles, i.e., $\theta_c = 90°$, and the left-most frame of Figure 6.13 is all we ever find. A monolayer of particles having no long-range particle interactions does not form a system, just a collection of components.

Let's now increase the friction among still-identical spherical particles, e.g., by using a material with a rough surface. A conical pile does form with $\theta_c < 90°$. The exact slope angle depends on the amount of friction (and other properties) introduced. Despite the introduction of long-range particle interactions through friction, the lack of component variability yields a homogeneous system. Avalanches occur but the homogeneity of the system gives these avalanches a *characteristic size*. Specifically, avalanche sizes will be normally distributed with mean and variance determined entirely by particle properties.

Now let the grain shapes vary, e.g., as we find in dry sand or rice, so there is significant particle-properties variability. Shape diversity enables the grains to interlock forming particle jams (Fig. 4.9) that may extend over a long spatial range. Repeating the experiment again, we discover a complex system has formed. We know this by virtue of the avalanches having no characteristic size. Properties that emerge at all scales will generate avalanches at all scales. This means that avalanche size S follows a *power-law distribution*.

This was the result found by [Bak *et al.* (1988)] (see Fig. 6.14) who simulated the sand pile experiment using a method known as *Cellular Automata*. The $1/S$ power-law distribution of avalanche sizes has no well-defined mean or variance, so if you were asked what is the average size of avalanches it would be difficult to answer. When plotted on log-log axes, this distribution appears as a straight line with a slope near -1. Often the power law is found over a finite range of the variable; more than two orders of magnitude of S in Figure 6.14. We can say there are many more small avalanches than large ones but there is no characteristic size that clearly reflects "average" particle properties. Since there are many eigenstates with local eigenvectors, it makes sense that larger avalanches are relatively rare events.

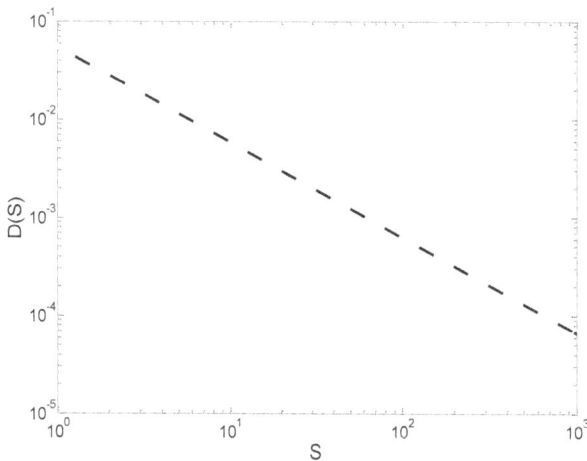

Fig. 6.14   Size of avalanches S in a 2-D simulation are plotted as a function of avalanche occurrence at that size $D(S)$ on a log-log scale. The results yield a power law $D(S) = S^m$ where in this case $m = -1$. Redrawn from the data in [Bak *et al.* (1988)].

In summary, simple/homogeneous and complex/heterogeneous systems can each achieve self-organized criticality. But only complex systems, which have the component variability necessary to create many degrees of freedom, will have power-law distributions. Many physical, biological, and man-made phenomena follow a power-law distribution, with or without criticality. These include the sizes of earthquakes known as Gutenberg and Richter Law [Gutenberg and Richter (1944)], waves in the ocean, solar flares, city populations, corporation size, income levels, and rainfall and drought durations. They also include the frequency of names in a region and words appearing in literature, known as Zipf's law [Zipf (1935)], as well as the topology of networks [Barabási and Albert (1999)]. Some of these cases that follow a power law distribution involve transitions in complex systems, such as earthquakes. Other cases, such as the occurrence of words in languages, follow a power law but are not related to any transitions. Behavior of the latter group follows Zipf's law. Our discussion in this section strictly applies to the former cases.

The power-law distributions of transitions reflect the high levels of component-property variability found in complex systems. As variability is lost, then avalanche mass becomes more characteristic of particle properties and therefore more predictable. This description applies at scales much larger than a sand pile in a lab. For example, it occurs when an earthen slope fails in the form of a mudslide or snow avalanches on a mountain side. If a mountain slope achieves a state of self-organized criticality, we can be sure at least several eigenvalues are near zero and susceptible to avalanche transitions, most likely small ones. If we set off a charge to make a slope safer for skiers, we are attempting to provide enough input energy to activate all eigenstates at least as large as any that might be activated during skiing conditions.

Perhaps the most interesting of all applications of self-organized criticality are in neuroscience. In 1999 Per Bak posed the idea that a normal functioning human brain is in a self-organized critical state. At first, neuroscientists were skeptical about this concept. However, gradually some started seriously considering this radical idea as the basis of a new way of looking into how the human brain functions. Now, this theory is being explored in the scientific community (e.g., see [Hesse and Gross (2014)]).

The brain operating at a self-organized critical state is similar to a sand pile that has reached a critical state. The size and distribution of electrical signals propagating through neural networks in the brain are analogous to that of avalanches occurring in a sand pile. That is, the normal brain appears to be in a state between two equilibrium states, one of quiescence and the other of chaos. It seems as brain activity goes subcritical or super-critical there are normal cellular mechanisms to return it to a critical state. The supposition is that evolution selected criticality as a trait for optimal neurological function. Neuroscientists have related disease conditions to a loss of criticality. Although in its infancy, the exploration of functional neural criticality appears ripe with possibility.

6.13   Summary

Component-property variability is fundamental to the formation of a complex system. Any large collection of widely-networked components can form a system; however, it is important to know the number of degrees of freedom in that system to understand if there will be a spatial hierarchy of eigenstates that give complex systems their unique emergent properties. Variability in component properties can increase the dimensionality of a system, which in modeling is the number of variables required to completely describe behavior. High-dimensional systems posses eigenstates with eigenvectors having a broad spectrum of spatial connectivity that gives complex systems their robustness to transitions. It is this robustness that enables complex systems to achieve a state of self-organized criticality. This is a state that places systems between stable equilibria so that frequent small-scale transitions place local properties in a constant flux, while the whole system remains very stable with respect to global properties. We believe this is the reason that variability in large systems can result in overall stability despite the appearance of local chaos.

Chapter 7

Epilogue

We wrote this book for readers interested in thinking about large complex problems and those pondering career paths likely to impact society's greatest challenges. In the course of writing, we discussed many topics with colleagues and advanced engineering students who raised a broad range of questions. How large must a system be to be classified as complex? How does one determine the intrinsic degrees of freedom for an arbitrary system? Can a high-dimensional system in dynamic equilibrium be prevented from undergoing what appears to be an inevitable transition to a disastrous equilibrium state? These are important questions about technical aspects that must be answered to develop effective systems analyses. We have our opinions as discussed here but no proven solutions. The ideas in this book were formed as the authors asked each other similar questions during weekly meetings taking place over years. The value of long-term collegial discussions for generating new ideas aimed at framing very complicated problems cannot be overstated.

Our discussions began as a formulation of Systems Taxonomy needed to discover core concepts underlying behavior. The goal was always to discover principles, and this text is a descriptive summary of what we learned along the way. We hope these ideas prompt others to undertake their own exploration. Answers will emerge once we learn to effectively monitor the stability of systems that are at best partially understood.

Another question we are frequently asked is, assuming eigenanalysis is a key tool, exactly how can eigenstates of large and highly-nonlinear systems be found? First, we posit that system eigenstates exist whether or not we have the tools to measure them directly. So finding methods for estimating uncoupled emergent properties of a system should be a priority if we hope to monitor and eventually guide future behavior.

One straightforward approach is to express the nonlinear system under investigation as a set of many differential equations that are instantaneously linearized at every instant in time. If the equations can be found, we could generate the Jacobian matrix linking input to output at each moment of time to watch the eigenstates evolve. The difficulty with this plan is that, even if we were able to obtain a complete mapping of every system detail, it would be obsolete before we could use it given the high rate that many complex systems adapt and evolve.

This effect is most clearly demonstrated in medicine with patients that are treated with drugs intended to block pernicious cell-signaling pathways (Section 5.10). It illustrates a fundamental limitation of the reductionist approach; there must be more systematic methods for discovering "indicators" or "biomarkers" or "test statistics" for large unknown complex systems. These are the names given to nonreductionist measures in economics, medicine, and related fields. They arise today mostly as *heuristic metrics* born from deep human experience. These summary measures attempt to serve the same role that eigenvalues serve in linear(ized) systems, and so we can appreciate their value. However, unlike the eigenstates computed for linear systems, heuristic metrics make no claim to independence or orthogonality. We need to think differently.

Although we cannot currently define the generalized eigenstates that describe uncoupled properties of systems, we see promising avenues for exploration. One approach is to first *mathematically model* key elements of each known component of a networked system. Then connect them in an *agent-based model* that facilitates component communication according to rules observed in natural systems. These computational approaches combine the strengths of mathematical modeling, to help us understand detailed mechanisms, with the heuristic rule-based models of experiments. These methods are being explored in various fields.

Systems-related challenges increasingly are pointing engineers toward the *data sciences* for answers. The central question is: How can a large pool of system-related information be systematically sampled and analyzed to accurately predict the response to input stimuli? It is clear that this question lies at the heart of many core problems in ecosystem management, economic planning, government strategies, and especially 21st century medicine. The problems we face in these areas are often overwhelming because we don't have a complete scientific understanding of how these enormous, complex systems respond and evolve over time.

We believe that *data-driven* computational modeling has a very bright future in systems analysis. Instead of using the physics of a situation to model processes mathematically, data-driven models use experimental measurements to learn how to predict system responses much in the same way expert humans learn. Often the methods employ neural-network structures and similar tools developed in computational-intelligence research. Already we can see that *machine-learning methods* are being successfully applied in areas such as *image recognition*. Under the label *Deep Learning* [Greenspan *et al.* (2016)], data-driven analyses are making significant progress in medical imaging and other computer-vision applications.

However, if this approach is to pay off generally, the fundamental principles governing the behavior of all systems must be firmly established and applied to guide machine-learning techniques. We also see a growing interest in applying systems-engineering principles to nontraditional areas involving natural and manmade systems; e.g., in economics, education, marketing, and social networking analyses. There is even a recent report on efforts to apply complex network theory to validate the historical accuracy of stories found in ancient literature [Kenna and MacCarron (2016)]. As we learn to adopt systems thinking into analysis, we find it has few, if any, disciplinary bounds. The challenges are great but the rewards for those who can successfully articulate system descriptors are enormous.

Bibliography

Ackerman, J. (2012). How bacteria in our bodies protect our health, *Scientific American* **June Issue**.

Alonso-Marroquin, F. (2008). Sheropolygons: a new method to simulate conservative and dissipative interactions between 2D complex shaped rigid bodies, *Europhysics Letters* **83**, 1, p. 14001.

Anderson, P. (1972). More is different, *Science* **177**, 4047, pp. 393 – 396.

Ashby, W. (1958). Requisite variety and implications for control of complex systems, *Cybernetica* **1**, pp. 83 – 99.

Attanasi, A., Cavagna, A., Castello, L. D., Giardina, I., Grigera, T., Jelic, A., Melillo, S., Parisi, L., Pohl, O., Shen, E., and Viale1, M. (2014). Information transfer and behavioural inertia in starling flocks, *Nat. Phys.* **10**, 9, pp. 691 – 696.

Bak, P. and Paczuski, M. (1995). Complexity, contingency, and criticality, *Proc. Natl. Acad. Sci. USA* **92**, pp. 6689 – 6696.

Bak, P., Tang, C., and Wiesenfeld, K. (1987). Self-organized criticality: An explanation of $1/f$ noise, *Phys. Rev. Lett.* **59**, 4, pp. 381 – 384.

Bak, P., Tang, C., and Wiesenfeld, K. (1988). Self-organized criticality, *Phys. Rev. A Gen. Phys.* **38**, 1, pp. 364 – 374.

Barabási, A.-L. and Albert, R. (1999). Emergence of scaling in random networks, *Science* **286**, 5439, pp. 509 – 512.

Bertalanffy, L. V. (1950). The theory of open systems in physics and biology, *Science* **111**, 2872, pp. 23 – 29.

Bertalanffy, L. V. (1951). Towards a physical theory of organic teleology: Feedback and dynamics, *Hum. Biol.* **23**, 4, pp. 346 – 361.

Bertalanffy, L. V. (1972). The history and status of general systems theory, *Acad. Manage.* **15**, 4, pp. 407 – 426.

Box, G. and Draper, N. (1987). *Empirical Model Building and Response Surfaces* (John Wiley & Sons, New York, NY).

Bush, V. (1960). *Science, the Endless Frontier; A Report to the President on a Program for Postwar Scientific Research* (National Science Foundation, Washington DC).

Crocket, C. (2015). How did Earth get its water? *Science News* **187**, 10, p. 18.

Darwin, C. (1859). *On the Origin of Species by Means of Natural Selection* (W. Clowes and Sons, London).

Dawkins, R. (1989). *The Selfish Gene, 2/e* (Oxford University Press, Oxford, UK).

Eldredge, N. and Gould, S. (1972). Punctuated equilibria: an alternative to phyletic gradualism, in T. Schopf (ed.), *Models in Paleobiology*, chap. 5 (Freeman, Cooper & Co, San Francisco), pp. 82 – 115.

Feynman, R., Leighton, R., and Sands, M. (1964). *The Feynman Lectures on Physics* (AddisonWesley, Boston MA, 3 volumes).

Gertner, J. (2012). *The Idea Factory: Bell Labs and the Great Age of American Innovation* (The Penguin Press, NY).

Ghaboussi, J. (2012). Unifying principles for sudden transitions in all systems, in *Proceedings of the Conference on Systems Engineering Research (CSER). Procedia Computer Science*, Vol. 8, pp. 75 – 80.

Ghaboussi, J. and Barbosa, R. (1990). Three-dimensional discrete element method for granular materials, *International Journal for Numerical and Analytical Methods in Geomechanics* **14**, pp. 451 – 472.

Ghaboussi, J. and Wu, X. S. (2016). *Numerical Methods in Computational Mechanics*, CRC Press, Taylor Francis.

Gillespie, C. (1983). *A Professionalization of Science* (Doshisha University Press, Kyoto).

Gödel, K. (1962). *On Formally Undecidable Propositions of Principia Mathematica and Related Systems* (Dover, Mineola, NY).

Greenspan, H., van Ginneken, B., and Summers, R. (2016). Deep learning in medical imaging: Overview and future promise of an exciting new technique, *IEEE Transactions on Medical Imaging* **35**, 5, pp. 1153 – 1159.

Gunderson, L., Allen, C., and Holling, C. (2010). *Foundations of Ecological Resilience* (Island Press, Washington DC).

Gutenberg, R. and Richter, C. (1944). Frequency of earthquakes in California, *Bulletin of the Seismological Society of America* **34**, pp. 185 – 188.

Hanahan, D. and Weinberg, R. (2011). Hallmarks of cancer: the next generation, *Cell* **144**, 5, pp. 646 – 674.

Hesse, J. and Gross, T. (2014). Self-organized criticality as a fundamental property of neural systems, *Frontiers Systems Neuroscience* **8**, 166, pp. 1 – 14.

Holland, J. (1995). *Hidden Order: How Adaptation Builds Complexity* (Helix Books, NY).

Holling, C. (1996). Engineering resilience versus ecological resilience, in P. C. Schulze (ed.), *Engineering Within Ecological Constraints* (National Academy Press, Washington DC), pp. 31 – 44.

Ingber, D. (2008). Tensegrity-based mechanosensing from macro to micro, *Prog. Biophys. Mol. Biol.* **97**, 2-3, p. 163179.

Johnson, R., Evans, J., Robinson, G., and Berenbaum, M. (2009). Changes in transcript abundance relating to colony collapse disorder in honey bees (Apis mellifera), *Proc Natl Acad Sci U S A* **106**, 35, pp. 14790 – 14795.

Kenna, R. and MacCarron, P. (2016). Maths meets myths, *Physics World* **29**, 6, pp. 22 – 27.

Lodish, H., Berk, A., Matsudaira, P., Kaiser, C., Krieger, M., Scott, M., Zipursky, S., and Darnell, J. (2004). *Molecular Cell Biology, 5/e* (W.H. Freeman and Co., New York).

Murray, J. (2002). *Mathematical Biology I: An Introduction* (Springer, NY).

NASA (2004). https://commons.wikimedia.org/wiki/File:Quake_epicen ters_1963-98.png.

Oberhardt, M., Palsson, B., and Papin, J. (2009). Applications of genome-scale metabolic reconstructions, *Mol Syst Biol* **5**, 320, pp. 1–15.

Oort, J. (1950). The structure of the cloud of comets surrounding the solar system and a hypothesis concerning its origin, *Bull Astron Inst Neth* **11**, pp. 91–110.

Palsson, B. (2006). *Systems Biology: Properties of Reconstructed Networks* (Cambridge University Press, New York, NY).

PCAST (2014). https://www.whitehouse.gov/sites/default/files/micro sites/ostp/PCAST/pcast_systems_engineering_in_healthcare_-_may_ 2014.pdf.

Pecknold, D., Ghaboussi, J., and Healey, T. (1985). Snap-through and bifurcation in a simple structure, *Journal of Engineering Mechanics Division, ASCE* **111**, 7, pp. 909 – 922.

Rittschof, C. and Robinson, G. (2016). Behavioral genetic toolkits: Toward the evolutionary origins of complex phenotypes, *Curr Top Dev Biol* **119**, pp. 157 – 204.

Schilling, C., Letscher, D., and Palsson, B. (2000). Theory for the systemic definition of metabolic pathways and their use in interpreting metabolic function from a pathway-oriented perspective, *J. theor. Biol.* **203**, pp. 229 – 248.

Steven, L. (1989). *Buckminster Fuller's Universe: His Life and Work* (Basic Books, New York).

Stokes, D. (1997). *Pasteur's Quadrant: Basics Science and Technological Innovation* (Brookings Institution Press, Washington DC).

Thompson, E. (2007). The tulipmania: Fact or artifact, *Public Choice* **130**, 1-2, pp. 99 – 114.

Weisskopf, V. (1965). *In defence of high-energy physics. In: Nature of matter - purposes of high-energy physics*, chap. 5 (Brookhaven National Laboratory, BNL 888 (T-360)), pp. 54 – 56.

Whitehead, A. and Russell, A. (1927). *Principia Mathematica, 2/e* (Cambridge University Press, UK).

Yanai, I. and Lercher, M. (2016). *The Society of Genes* (Harvard University Press, Cambridge, MA USA).

Zachary, G. (1997). *Endless Frontier: Vannevar Bush, Engineer of the American Century* (Free Press, NY).

Zaug, A. and Cech, T. (1986). The intervening sequence RNA of tetrahymena is an enzyme, *Science* **231**, 4737, pp. 470 – 475.

Zhao, D. (2008). Real-time soil models for machine-medium interaction in virual reality, *PhD Thesis, Department of Civil and Environmental Engineering, University of Illinois at Urbana-Champaign*.

Zhao, D., Nezamiand, E., Hashash, Y., and Ghaboussi, J. (2006). Three-dimensional discrete element simulation for granular materials, *Engineering Computations* **23**, 7, pp. 749 – 770.

Zipf, G. (1935). *The Psycho-Biology of Language* (Houghton-Mifflin, Boston).

Index